Remote Sensing and GIS in Climatological Modelling

By

Sailesh Samanta

Department of Surveying and Land Studies
Papua New Guinea University of Technology
Lae, Morobe, Papua New Guinea

Preface

This book presents an introductory course of study in Remote Sensing and GIS and its application for climatological modelling. Geographic information systems (GIS) and modelling are becoming powerful tools in agricultural research and natural resource management. Spatial information about climate variables is very important for a sound agricultural planning. Many disciplines use these climate variables (temperature, precipitation, atmospheric pressure, cloudiness and humidity) as a basis to understand the processes they study. Evidently the information is limited to the available meteorological stations and therefore to discrete points of the space. To understand the weather and climatic condition over a continuous space (where weather or climate station is not present) one needs to resort to the modern techniques like spatial interpolation by GIS. So far the work on climatological interpolation dovetailed to themes like 'commercial agriculture' has been scanty.

This book is focused on statistical model to forecast the temperature and the precipitation using remote sensing and geographic information system (GIS) techniques. It also examines statistical approaches for interpolating climatic data over large regions, providing different interpolation techniques for climate variables for use in agricultural research.

Gangetic West Bengal is a region of strong heterogeneous surface with several weather disturbances. Different climate variables like average maximum, mean, minimum temperature and rainfall are used for the interpolation process as the dependent parameters and other variables like relative humidity and cloudiness factor are used as the independent parameters for temperature and rainfall modelling. Standard false color composition bands (green, blue and near-infrared bands) of LANDSAT-7, ETM+ are used to produce land use/land cover dataset. Digital elevation model is built using the contours which are pulled together from Topographical maps of the region and SRTM data sets. With the help of soil type map of West Bengal and soil region map of India the soil texture dataset is built. Land use/land cover, soil texture and digital elevation model are used as the independent variables for temperature modelling. Rainfall modelling is performed using relative humidity, cloudiness factor, latitude, land use/land cover and digital elevation model. Multiple regression analysis with standard method is used to add dependent variables (temperature and rainfall) into regression equation. Finally standard deviation errors are evaluated after comparing the predicted and observed temperature and rainfall of the area. Interpolated dependent climate variables, dependent parameters, the model output and the standard error raster matrixes are used to build final maps applying basic GIS techniques.

The book is being laid out in Eight (8) chapters.

Chapter 1 deals with the general introduction of weather and climate, significance of the study, objectives and layout of the study.

Chapter 2 discusses the geographical setting of the study area. This chapter deals with precise geographical locations of the study area, topography, geology, drainage, climate, soil, vegetation and physiographic setting of the study, also covers a brief statement about the relevant literature review.

Chapter 3 deliberates on the study of land use /land cover characteristics and soil texture. This chapter also discusses certain basics of remote sensing processes to determine land use/land cover and soil texture of the study area in different scale. It also discusses influence of land cover and soil texture on weather.

Chapter 4 deals with topographic characteristics of the study area, surface analysis using digital elevation model and its importance on temperature and rainfall variation.

Chapter 5 discusses land surface temperature assessment using thermal bands of remote sensing data (LANDSAT 7, ETM+).

Chapter 6 deliberates on spatial interpolation of dependent climate variables and climatological modelling of temperature and rainfall using various dependent and independent climate variables in the approach of neural network modelling. This chapter also deals with the results and verification analysis of temperature and rainfall modelling.

Chapter 7 discusses modelling of temperature and rainfall in West Bengal.

Chapter 8 contains the conclusions and recommendation of the total study carried out.

I take this privilege to express my heartfelt gratefulness to Dr. Dilip Kumar Pal, Professor and Head, Department of Surveying and Land Studies, Papua New Guinea University of Technology for his continuous encouragement and able guidance throughout the length of the research.

I wish to express my heartiest thanks to Dr. Debasish Lohar, Reader, Atmospheric Science Group, Department of Physics, Jadavpur University for his talented and inspiring guidance and constant encouragement throughout the course of research.

I would like to express my special thanks to Dr. Jatisankar Bondhopadhya and Mr. Abhishek Choakroborty, Department of Remote Sensing and GIS, Vidyasagar University, for their moral support and suggestions during the course of study.

I would like to express my sincere thanks to Dr. Ashis Kumar Paul, Dr. Ramkrishna Maiti, Dr. Soumendu Chatterjee, Mr. Utpal Roy and Dr. Nilanjana Das, Faculties of Department of Geography and Environment Management, Vidyasagar University, for their valuable and timely suggestions to improve my research work during the study.

My thanks go to Mr Pravat Shit and Mr Parimal Sarkar, Staff of the Department of Geography and Environment Management, and Mr Sumanta Saha, staff of the Department of Remote Sensing and GIS Vidyasagar University, who helped directly or indirectly in the completion of my research work.

Special thanks go to Animesh Majhi, Rajkumar Giri, Tithi Sarkar, Basushree Girimohanto, Budhodev Mondal, Krishanu Achariya, Sudha Krishana Mondal, Sudin Giri, Jahir Khan, Arindam Pal, Milan Das and Palash Das, Ex-students, Department of Remote Sensing and GIS for their help to collect soil data from different location of the study area during the period of research.

I also convey my gratitude to parents and my elder brother, who are constant source of inspiration throughout my life and without whom I could not have reached this stage.

Last, but not the least I would like to pay my heartiest and special thanks to my beloved wife, Babita Pal for her technical support and co-operation in S/w based work, moral support, suggestions and affection right from the day one to this day of completion of my research work.

List of Figures

List of Tables

List of Appendixes

List of Abbreviations

AML -Arc Macro Language

ANN -Artificial Neural Networks

AOI -Area of Interest

ASCII -American Standard Code for Information Interchange

CRU -Climatic Research Unit

DEM -Digital Elevation Models

DN -Digital Number

EGM96-Earth Gravitational Model 1996

ERDAS-Earth Resources Data Analysis System

ESRI -Environmental Systems Research Institute

ETD -Extra-tropical disturbances

ETM$^+$ -Enhance Thematic Mapper Plus

FCC -False color composition

GCP -Ground control point

GIS -Geographic Information System

GSI -Geological Survey of India

GT -Ground Truth

IDW -Inverse Distance Weighted

IDWA -Inverse distance weighted averaging

IMD -India Meteorological Department

ITCZ -Intertropical Convergence Zone

KTPP -Kolaghat Thermal Power Plant

KTPS -Kolaghat Thermal Power Station

LANDSAT-Landsat Satellite

LIDAR-Light Detection and Ranging

LSM -Land Surface Model

LST -Land Surface Temperature

LUT -Look up' Table

MWR -Monthly Weather Rewiew

NATMO-National Thematic Mapping Organization

NBSS -National Bureau of soil survey

NCDC -National Climate Data Center

NDVI -Normalized Difference Vegetation Index

NH -National Highway

NRSA -National Aeronautics and Space Administration

RADAR-Radio Detection and Ranging

RGB -Red Green Blue

RMS -Rout Mean Square

SOI -Survey of India

SRTM -Shuttle Radar Topography Mission

SSE -Surface meteorology and Solar Energy

SWBD-Shorelines and Water Bodies Database

TIN -Triangulated Irregular Network

TM -Thematic Mapper

USDA -United States Department of Agriculture

USGS -United State Geological Survey

WGS 84-World Geodetic System 1984

Contents

CHAPTER 1

INTRODUCTION

CHAPTER OUTLINE

- The elements of climate
- Climate classification
- Significance of the study
- Objectives of the study

1.0. INTRODUCTION

The Earth's climate is generally defined as the average weather over a long period of time. A place or region's climate is determined by both natural and anthropogenic (human-induced) factors. The natural elements include the atmosphere, geo-sphere, hydrosphere, and biosphere, while the human factors comprise land and other resource uses / abuses emanating from demographic and social dynamism. Changes in any of these factors can cause local, regional, or even global changes in the climate. Weather is the current atmospheric conditions contributed by temperature, rainfall, wind, and humidity at a given place. It is what we see or feel in the ambience i.e. rain or wind, or sunny or cloudy or appraise how hot it is by taking a temperature reading. So weather is the current state of the ambience / atmosphere. On the other hand, climate is the general weather conditions over a long period of time (Henderson et al. 1993). Some meteorologists aptly simplify "climate is what you expect and weather is what you obtain". According to one middle school student, "climate dictates what clothes to buy, but weather tells what clothes to wear" (Henderson et al. 1993). Climate is sometimes referred to as "average" weather for a given area. The national weather service uses data for instance temperature as high and low and precipitation rates for the past thirty years to compile an area's "average" weather. However, some atmospheric scientists think that you need more than "average" weather to accurately portray an area's of climatic character - variations, patterns, and extremes must also be included (Henderson et al. 1993). Thus, climate is the sum of all statistical weather information that helps to describe a place or region. The term also applies to large-scale weather patterns in time or space such as an 'ice age' climate or a 'tropical' climate.

It is evident that an area's climate keeps changing, although changes do not usually occur on a time scale that's immediately obvious to us. While we know how the weather changes from day to day, subtle climate changes are not as readily detectable. Weather patterns and climate types take similar elements into account, the most important of which are (i) the temperature of the air, (ii) the humidity of the air, (iii) the type and amount of cloudiness, (iv) the type and amount of precipitation, (v) air pressure and (vi) wind speed and direction. Although weather and climate are dissimilar but they are very much interrelated. Changing of one weather variable affects on others simultaneously in the regional climate. For example, if the average temperature over a region increases significantly, it can influence the amount of cloudiness as well as the type and amount of precipitation that occur. If these changes occur over long periods of time, the average climate values for these elements will also be affected.

1.1 The elements of climate

Climatology is the study of the long-term state of the atmosphere, or climate. The long-term state of the atmosphere is a function of a variety of interacting elements. They are (i) solar radiation, (ii) air masses, (iii) pressure systems (and cyclone belts), (iv) ocean currents and (v) topography. Especially the first three elements are contingent on the latitudinal setting of the area.

Solar radiation is probably the most important element of climate and is aptly regarded as the 'driver' of climate (Ritter 2006). Solar radiation predominantly heats up the Earth's surface which in turn determines the temperature of the air above, albeit moderated by the composition and concentration of greenhouse molecules overlying. The latter entails a lower surface temperature of high altitude terrain. The incoming solar radiation along with wind drives evaporation, so long as water is available. Under a given irradiation condition, the extent of evaporation is also dependent on the nature of water available viz. inland fresh surface water, sea water, or soil moisture. Accordingly the humidity of a given area is contingent on the irradiation, availability and nature of water source, and the prevailing wind condition. Heating of the air above ground determines its stability along the profile, the instability triggers rise of moisture laden air followed by cooling of water vapour culminating in cloud development and precipitation. Unequal heating of the Earth's surface creates pressure gradients that result in wind. So it is evident that almost all the characteristics of climate can be traced back to the receipt of solar radiation as the driving force.

Air masses as an element of climate subsume the characteristics of temperature, humidity, and stability (Ritter 2006). Location relative to source regions of air masses in part determines the variation of the day-to-day weather and long-term climate of a place. For instance, the stormy climate of the mid-latitude owes its existence to vicinity of the boundary zone of greatly contrasting air masses called the polar front.

Pressure systems have a direct impact on the precipitation characteristics of different climate regions. In general, places dominated by low pressure tend to be moist, while those dominated by high pressure are dry. The seasonality of precipitation is affected by the seasonal movement of global and regional pressure systems. Climates located at 10° to 15° of latitude experience a significant wet period when dominated by the Inter-tropical convergence zone and a dry period when the subtropical high carves into this region as a temporal cyclical swing. Likewise, the climate of Asia is impacted by the annual fluctuation of wind direction due to the monsoon. Pressure dominance also affects the receipt of solar radiation. Places dominated by high pressure tend to lack cloud cover and hence receive significant amounts of sunshine, especially in the low latitudes (Ritter 2006).

Ocean currents greatly affect the temperature, humidity and precipitation of a region. Those climates bordering cold currents tend to be drier as the cold ocean water helps stabilize the air by precluding upwelling to inhibit cloud formation and precipitation. Air traveling over Cold Ocean currents lose energy to the water and thus moderate the temperature of nearby tropical coastal locations. Air masses traveling over warm ocean currents are replete with moisture to promote instability and precipitation. Additionally, during winter the high thermal inertia of warm ocean water with the help of global and local wind circulation keeps air temperatures somewhat warmer than locations inland and away from the coast as evidenced in Canada, Western Europe and Eastern Europe.

Topography affects climate in a variety of ways. The orientation of mountains to the prevailing wind affects precipitation. Windward slopes, those facing into the moisture laden wind, experience more precipitation due to orogenic uplift of the air along the slope. Leeward sides of mountains are in the rain shadow and thus receive less precipitation owing to the depleted moisture content of the wind. Air temperatures are affected by slope and orientation as slopes facing into the Sun will be warmer than those facing away. Temperature also decreases as one move toward higher elevations. Mountains have nearly the same effect as latitude does on climate. On tall mountains a zonation of climate occurs as we move towards higher elevation.

1.2 Climate classification

The purpose of climate classification is to organize a set of data or information about climate to effectively communicate it in an informative way. Climate classification helps synthesize information into smaller units that are more easily understood. When considering the Earth's climate, there is such an enormous amount of information that one has to break it down into areas of commonality to easily understand it. Climatologists have therefore created several ways to organize the wealth of information about Earth's climate to bring order and understanding to it. There are three fundamental types of classifications used in climatology. First there are empirical systems of classification that are based on observable features. Genetic classification systems are those based on the cause of the climate. Applied classification systems are those created for, or as an outgrowth of, a particular climate-associated problem. The Koeppen system discussed below is an empirical system based on observations of temperature and precipitation (Kottek et al. 2006). These are two of the easiest climate characteristics that can be measured, and probably the ones with the longest historical record. It's fairly easy to collect air temperature readings with a thermometer and precipitation with some sort of collecting device that can measure the amount of precipitation. Climates are grouped based on annual averages and seasonal extremes.

The entire study area (21° N - 25° N and 85° E - 89° E) falls under the "Aw" (equatorial-winter dry) climate. According to the Koeppen classification system the study area is covered under the main climate as equatorial ("A") and under precipitation as dry winter.

15

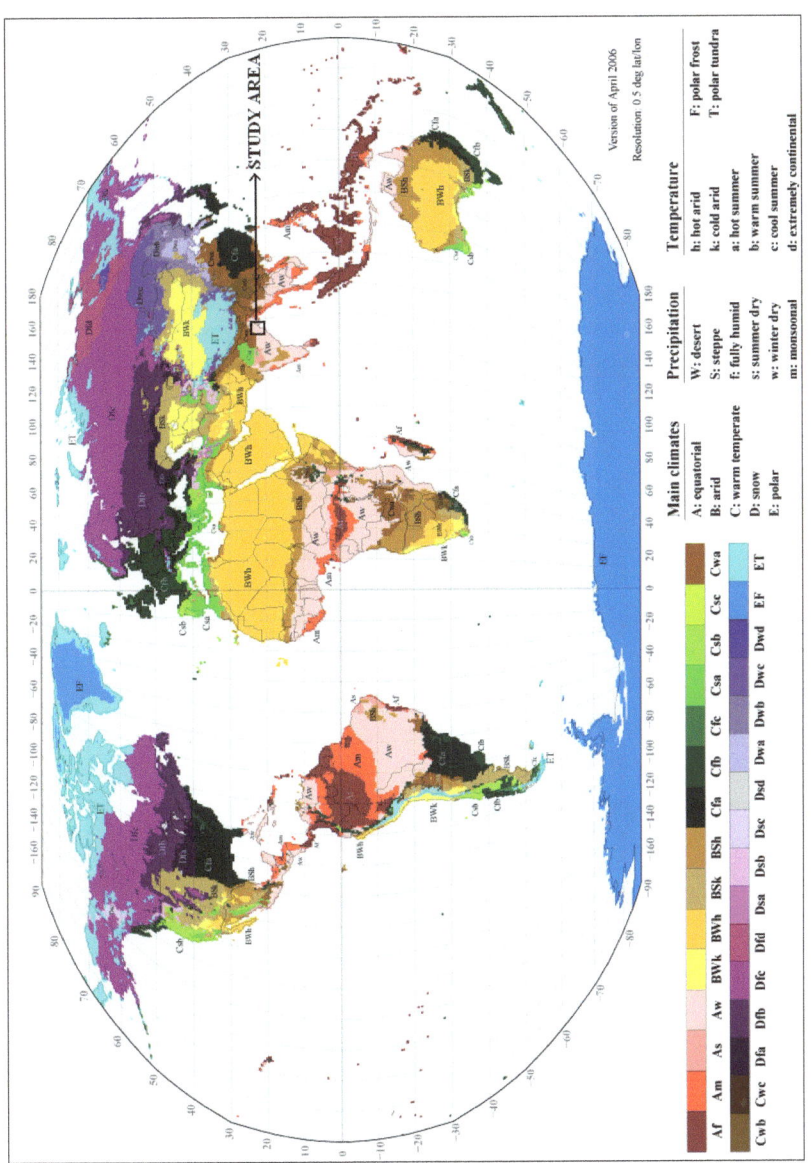

Figure 1.1. World climate patterns according to Koeppen-Geiger, updated with CRU TS 2.1 data set (1952 to 2000)

Source: http://gpcc.dwd.de;   http://koeppen-geiger.vu-wien.ac.at

16

1.3 Significance of the study

Agriculture in India is the means of livelihood of almost two thirds of the work force in the country. Although in terms of its share to India's GDP it ranks third, yet it has always been India's most important economic sector in terms food security and job creation. Climate and agriculture are interrelated entities, both having global implications. Global warming is projected to have significant impacts on conditions affecting agriculture, emanating from a perilous regime of 'temperature, 'ambient carbon dioxide concentration', 'precipitation' and 'the mutual interaction of these elements'. These conditions determine the carrying capacity of the biosphere to produce enough food for the burgeoning human population and domesticated animals (Fraser 2008). Geographic information systems (GIS) and modelling are becoming powerful tools in agricultural research and natural resource management. Spatial information about climate variables is very important for a sound agricultural planning. Many disciplines use these climate variables (temperature, precipitation, atmospheric pressure, cloudiness and humidity) as a basis to understand the processes they study. Evidently the information is limited to the available meteorological stations and therefore to discrete points of the space. To understand the weather and climatic condition over a continuous space (where weather or climate station is not present) one needs to resort to the modern techniques like spatial interpolation by GIS. So far the work on climatological interpolation dovetailed to themes like 'commercial agriculture' has been scanty. Traditionally, the applied method had been the lineal interpolation between stations approximated to the drawing of isolines based in the researcher knowledge of the studied area. There are recent works that search statistical relationships between geographical variables and climatological as well as studies that use GIS (geographical information systems) to modulate these climatological variables (Miquel et al. 2000). This work attempts to develop statistical model to forecast the temperature and the precipitation (monthly and annual) in the eastern part of India (21° N - 25° N and 85° E - 89° E).

1.4 Objectives of the study

The work has been undertaken with the intent of addressing the following nine objectives.

- To find out land use/ land cover pattern of the area using satellite images.
- To prepare soil texture map of the area.
- To build up digital elevation model of the area.

- To interpolate the existing climate variables to distribute in the total study area.

- To predict monthly and annual temperature through statistical analyses.

- To predict total monthly and annual rainfall through statistical techniques.

- To carry out a multiple regression analysis between predicted and existing result.

- To find out the corrector using the predicted and existing result.

- To carry out mapping of above prediction.

CHAPTER 2

GEOGRAPHICAL SETTING AND REVIEW OF LITERATURE

CHAPTER OUTLINE

- Geographical setting
- Physiographic setting of the study area
- Review of literature

2.0 GEOGRAPHICAL SETTING AND REVIEW OF LITERATURE

2.1 Geographical setting

2.1.1 Location

The study area is extended from 21° N to 25° N and 85° E to 89° E (Figure 2.1). Major portion of eastern India and western part of Bangladesh come under the above longitudinal and latitudinal extension. The study area includes most of the province of West Bengal, eastern part of Jharkhand, north-east part of Orissa and a small southern part of Bihar in India. The study area is embodied by survey of India topographical map numbers 72G, 72H, 72K, 72L, 72O, 72P, 73E, 73F, 73I, 73J, 73M, 73N, 78C, 78D, 79A and 79B and India and Pakistan AMS topographic maps (U.S. army map service) numbers NG-45/13 to NG-45/16 and NF-45/1 to NF-45/12. Also the study area corresponds to the LANDSAT-7 ETM+ imageries with path row numbers 138_43, 138_44, 138_45, 139_43, 139_44, 139_45, 140_43, 140_44 and 140_45.

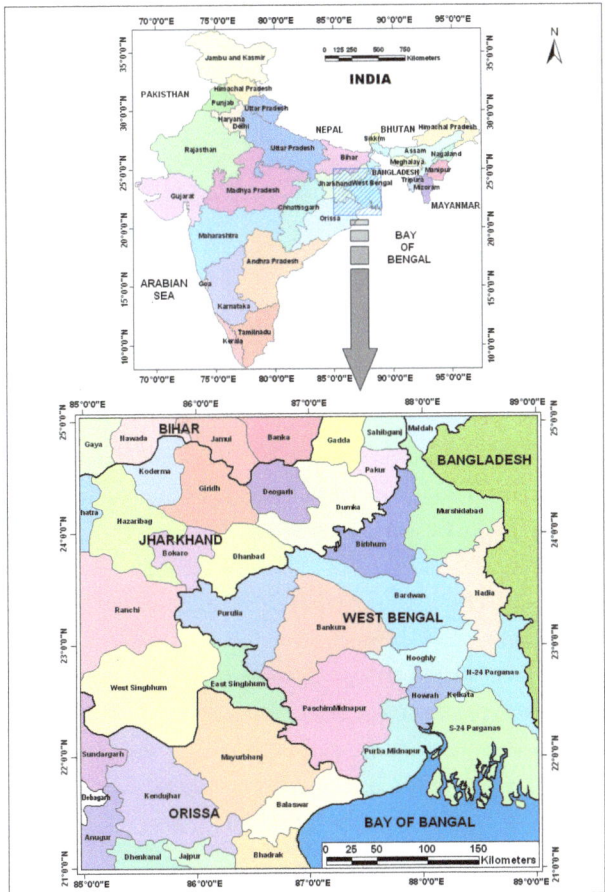

Figure 2.1. The location map of the study area

2.1.2 Topography

The southern region of the province of West Bengal, popularly known as South Bengal is demarcated by the river Ganges flowing east-west along the boundaries of the districts of Malda on the north and Murshidabad in the south, the latter being the northernmost district of South Bengal. South Bengal consists of two geographically distinct areas viz. "the western plateau fringe" and "the vast alluvial plain" of east and southeast. "Western plateau fringe" consists of the entire Purulia district and the western part of four other districts of West

Bengal viz. Birbhum, Burdwan, Bankura and Midnapur. The highest point of this plateau, named Goraburu Hill in Purulia district is 677 m and the lowest point is 85 m above the sea level where the up land ends in Midnapur district on the northern bank of the Subarnarekha River the altitude falls to 50 m above the sea level. The 'plateau fringe' forms the tail-end of the Chhotonagpur plateau that spreads toward west in the Jharkhand state and the Chhattisgarh state carved from erstwhile Orissa and Madhya Pradesh provinces. The rest of the east and southeast region is a vast alluvial plain. The vast alluvial plains of the West Bengal spread from Jalpaiguri and Siliguri in the north to the Sundarban creeks and its Kanthi littoral in the south. The coastal fringe likewise is of two distinct characters, west of the Hooghly the coastal strip in Midnapur district called the Contai or Kanthi strip. It consists of sand dunes and salt marshes mingled with each other. The marshes are formed behind well-developed sand bars. At places there are large shifting sand dunes, which have a tendency to blow landwards and encroach upon the cultivated land behind them. Vistas of casuarinas plantation are being developed all along the coast to fix the dunes and stop sea erosion. The Japanese quick growing creeper Kudzu is also being planted.

On the basis of homogeneity, continuity and physiographical characteristics, the state of Orissa has been divided into five major morphological regions, the Orissa coastal plain in the east, the middle mountainous and highlands region, the central plateaus, the western rolling uplands and the major flood plains. The Orissa coastal plains are the depositional landforms of recent origin and geologically belong to the post-tertiary period. The 75 m contour line delimits their western boundary and differentiates them from the middle mountainous region. This region stretches from the West Bengal border, i.e. from the River Subarnarekha in the north to the River Rushikulya in the south.

The middle mountainous and highlands region cover about three-fourth of the entire state. Geologically it is a part of the Indian Peninsula which in turn is a part of the ancient landmass of the Gondwanaland. The major rivers of Orissa with their tributaries have cut deep and narrow valleys. This region mostly comprises the hills and mountains of the Eastern Ghats which rise abruptly and steeply in the east and slope gently to a dissected plateau in the west running from north-east (Mayurbhanj) to north-west (Malkangirig). This region is well marked by a number of interfluves or watersheds. The Eastern Ghats is interrupted by a number of broad and narrow river valleys and flood plains. The average height of this region is about 900 m above the mean seal level. The plateaus are mostly eroded plateaus forming the western slopes of the Eastern Ghats with elevation varying from 305 m to 610 m. There are two broad plateaus in Orissa: (i) the Panposh - Keonjhar -Pallahara plateau comprises the upper Baitarani catchment basin, and (ii) the Nabrangpur - Jeypore plateau comprises the Sabari basin. The western rolling uplands are lower in elevation than the plateaus having heights varying from 153 m to 305 m. The elevation of the eastern and southern parts of the study area are less than 100 m while the western and south-western parts are relatively more elevated (Figure 2.2).

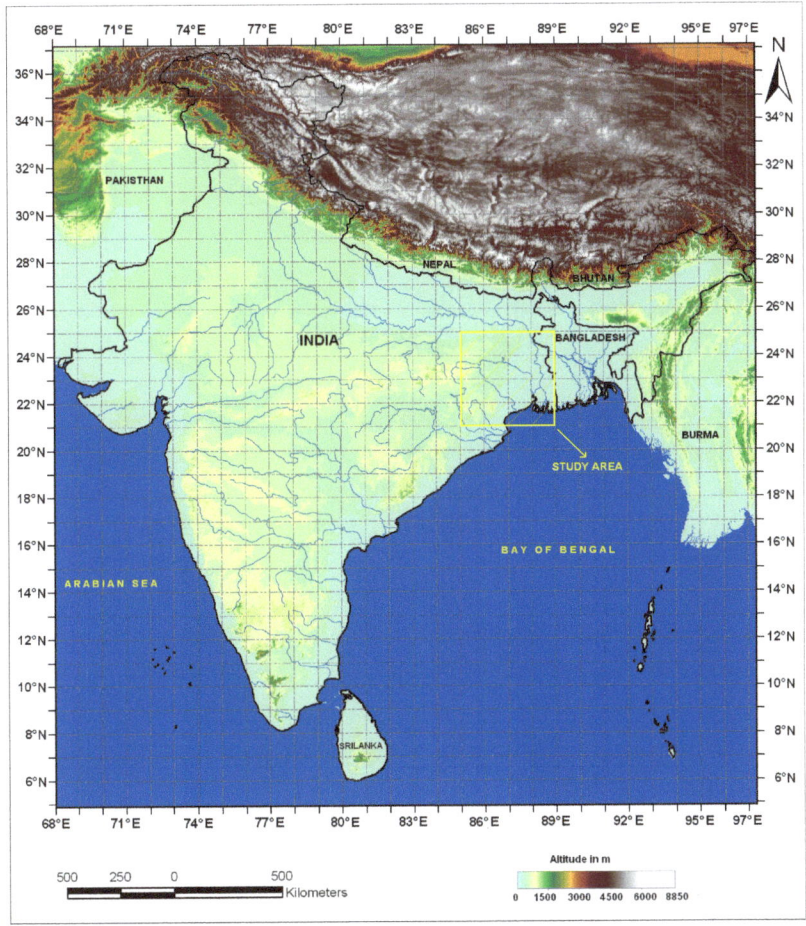

Figure 2.2. General elevation map of India based on topographic maps and SRTM data

2.1.3 Geology

Geologically the study area is a part of the Indian Peninsula which is also a piece of the ancient landmass of the Gondwanaland. The hill are principally composed of hard grey and grayish white gritty quartzite's, associated with large masses of irregular veins of quartz; as a whole the rocks are much twisted and contorted. Bihar plateau province roughly covers the south-eastern part of the state of Bihar and the western fringe of West Bengal. The entire area manifests in a plateau type of landform with occasional hills. The plateau is sloppy eastward from an altitude of 600 m around Ranchi to 178 m around Purulia and gradually merges through the laterite rim to the coastal plain province. The topography changes gradually from rugged and highly undulating at higher altitudes to slightly undulating or rolling at lower elevation. In general, the province occupied by rocks of pre-quaternary age viz. Precambrian, Gondwana sedimentary rocks, Rajmahal basalts and upper tertiary sediments. Moderate to strongly weathered tropical soils, lateritic soil and laterites have developed on these older rocks. The Singbhum area of south Bihar where three Precambrian thrust zones running roughly east west for about 150 km before taking a southerly bend immediately to the southwest of Bengal basin (Dunn 1941). Chhotanagpur area is characterised by large spread of granite gneiss. A tertiary gravel and grit (Dunn 1941) are found to occur between the Precambrian and quaternary alluvium in Midnapur and Bankura districts of West Bengal. These beds have also been laterized at many places. Lateritic soils, laterites and other subtropical weathered soils under study at lower altitude are within the area of Pleistocene age. This so-called 'low level laterites (or soils) are considered by Indian geologists (Krishnan 1968; Pascoe 1973) to be a secondary laterite derived by erosion of the adjacent 'high level laterite' formed high plateau to the west. Though this view has been contradicted by Niyogi (1970), where they described some of the laterite profiles on rocks and unconsolidated materials of various ages. Chhotanagpur plateau province and its adjoining portion of the coastal plain province because of its geomorphic antiquity show more relief (usually more than 50 m) and a much greater degree of dissection than the quaternary terrain in the east.

2.1.4 Drainage

The southern region of West Bengal is bisected by the Bhagirathi (Hooghly) river (Figure 2.3), one of the two forks of the Ganga that take off from top of Murshidabad district. The plain land on the western bank of the Hooghly River is largely formed by the deposits carried by a system of Hill Rivers rising in the western hills that pour their waters into the Hooghly and form part of the Gangetic delta. The main river in this system is the Damodar, Bengal's 'River of Sorrow'. The plains to the east are watered by distributaries of the Ganga branching off in West Bengal as well as Bangladesh. One feature of these plains is the existence of shallow lagoons called 'dahas'or 'boonrs', formed by beds of distributaries that got silted up above and below and of low marsh lands called bells that become flooded during the rainy season.

In Bihar Ganga is the main river which is joined by tributaries with their sources in the Himalayas (Elhance 1999). Some of them are Saryu (Ghaghra), Gandak, Budhi Gandak, Bagmati, Kamla-Balan and Mahananda. There are some other rivers that start from the plateau area and meet in Ganges or its associate rivers after flowing towards north. Some of them are Sone, Uttari Koyal, Punpun, Panchane and Karmnasha.

Rivers in Orissa though not perennial, serve the basic requirement of the state. All the rivers of Puri distict have common characteristics. In the hot weather they are beds of sand with tiny streams or none at all, while in the rains they receive more water than they can carry. The situation is conducive to rice-agriculture and is responsible for the state being an agrarian-economic. Main rivers of the Orissa are: the Subarnarekha, the Brahmani and the Mahanadi.

Mahanadi is the major river of Orissa and the sixth largest river in India. It originates from the Amarkantak hills of the Bastar Plateau in Raipur district of Madhya Pradesh. It is about 857 km long (494 km in Orissa) and its catchment area spreads over 141,600 sq km. (65,580 sq km in Orissa). The river carries on an average about 92,600 million m^3 of water. The Brahmani, the second largest river in Orissa is 799 km Long.

The Subarnarekha comes in from Jharkhand and goes out to Orissa before it flows into the Bay of Bengal. It traverses across the south-west of Medinipur district along an arc like curved course. It mainly flows through Jhargram subdivision. It is said that traces of gold are found on its bed and as such it has derived the name (Subarno = gold, rekha = line) in local parlance. But the prohibitive cost of recovery of gold owing to very low concentration has been discouraging the foragers of tracer gold of late. The river has a shallow and narrow thalweg though the sandy bed, which is quite wide at places (as wide as ¾ km or more).

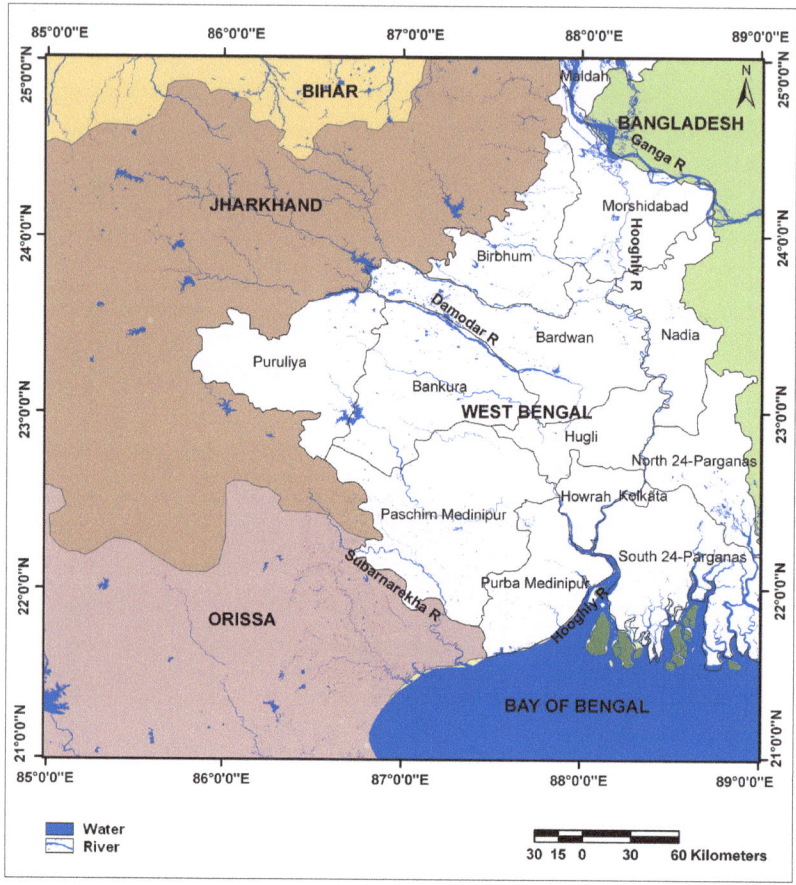

Figure 2.3. Drainage map of the study area based on satellite image and topographical maps

2.1.5 Climate

The climate of West Bengal is full of variation. The area experiences a tropical type of climate. The seasons of the West Bengal can be broadly categorized into four seasons (Das et al. 2011): winter (December – February), premonsoon (March through May), monsoon (June through September) and postmonsoon (October – November). The day temperature in premonsoon season ranges from 38° C to 45° C in different parts of the state. At nights, cool southerly breeze carrying moisture from the Bay of Bengal is usually present. The high temperature often causes troughs of low pressure to form on the plains and is compensated by sudden spell of brief storms known as kal-baisakhi or 'nor-westers', made of thunder showers to usher in a bout of relief from scorching heat. These summer storms can be quite destructive but the rainfall associated with these storms are very much useful for this part of the country as well as in Bangladesh (Ahmed 1989). The monsoon normally arrives by the middle of June. It scouts arrival about two weeks before its normal onset, which is characterized by somewhat uninterrupted phase of downpour with thick overcast prevailing over months. The monsoon rains in West Bengal are caused solely by the current of moisture laden wind from the Bay of Bengal. Variability is a characteristic feature of the monsoon in West Bengal as well as Bangladesh and Orissa which all receive the impact of the south-west Bay current. Breaks in the continuity of rain are not unusual; the resultant thoughts of low pressure develop into cyclonic storms especially in the postmonsoon season. A welcome change in the weather begins to be distinctly felt towards the end of September. Towards the end of monsoon season and Postmonsoon season in West Bengal is the season for festivity in the fields when the golden grain of paddy starts ripening and is harvested towards the end of the season. The conclusion of the round of the festivities marks the onset of the winter in November. Winter, which lasts about three months, is mild over the plains, the average minimum temperature not usually falling below 15° C. It is accompanied by a cold and dry northern wind, substantially lowering the humidity level. The weather gets warmer by the middle February, which heralds a brief spring season lasting about a month during which the deciduous trees break out in young green leaves and flowers. But this mellow season is too short-lived and the heat is turned on until with the coming of April, clammy summer comes in full blast and the annual cycle of seasons rolls on once again. Orissa also enjoys a tropical monsoon type of climate like most other parts of the study area. Its annual average rainfall is about 200 cm. The south-west monsoon normally sets in between 5th June in the coastal plain, and by 1st July the whole of the northern part of the country is under the full sway of the south-west monsoon (Rao 1981). By 15th October, the south-west monsoon withdraws completely from Orissa. Orissa, on the eastern coast of India, is not directly influenced by south-west monsoon branch from the Bay of Bengal like the west coast from Arabian sea and Indian ocean, but the annual cyclones from the Bay of Bengal influence it and bring copious rain with two seasonal peaks, July-August and October-November. During the winter, except northern Orissa, all other parts remain almost dry. Monsoon rainfall is of highest importance as it directly controls the crop condition in Orissa. Monsoon rainfall is considered 'high' when it is above 140 cm and 'low' when it is less than 120 cm as this is barely sufficient for a good paddy harvest. Winter rainfall helps the growth of the second crop in Orissa and hence is of importance. A total rainfall of above 5 cm during the winter has been taken to be 'high' while less than 2.5 cm is 'low' in Orissa. Flood and drought are common to many states in India, but only a few states are subject to cyclones. Orissa is one of the few. Moreover cyclone frequency also shows an increasing trend (Singh et al. 2001). The climate of Jharkhand is on the whole dry and bracing. The average annual rainfall for the state as a whole is 133.5 cm. From the onset of the monsoon by the middle of June, rainfall rapidly increases reaching the peak level in August. The annual variation of rainfall is not much. December and January are the coolest months. By March temperature begin to rise steadily. In May and early part of June the maximum temperature can be as high as 47° C on individual days. Humidity is generally normal in this district, except in monsoon months. The mean temperature in November for all over Bihar varies from 19.6° C to 22.2° C. The mean temperature in Gaya and Patna in December is 17° C and 18.2° C respectively. January is the coldest month in Bihar. The mean minimum temperature varies from 7.5° C to 10.5° C though some places like Netarhat record much lower temperatures than 7.5° C. The highest temperature is often registered in May which is the hottest month in the state. Like the rest of the northern India, Bihar also experiences dust-storms, thunder-storms and dust raising winds during the hot season. Dust storms having a velocity of 48 to 64 km/hour are most frequent in May and with second maximum in April and June. The hot winds (called 'Loo' in local parlance) of Bihar plains blow during April and May with an average velocity of 8 to 16 km/hour. This hot wind greatly affects human comfort during this season. The rainy season begins in June. The rainiest months are July and August. The rains are the gifts of the south west monsoon. There are three distinct areas in Bihar where rainfall exceeds 1800 mm. Two of them lie on northern and north-western wings of the state and the third lies in the Netarhat area. The south-west monsoon normally withdraws from Bihar in the first week of October (Rao 1981).

2.1.6 Soil

Soil is one of the most important resources of a nation. It is the priceless gift of nature. The most common use of the word soil is in the sense of a medium in which plants grow, although it has a different connotation at different time and place, and for persons engaged in different professions. Almost all the economic activities are directly or indirectly dependent on soil. Thus soil is the backbone of agricultural and industrial development. Soil has a number of characteristics, which may be regarded as the aggregate of the physical, chemical and biological properties. The Bihar plane consists of a very thick alluvial mantle of drift origin overlying in most part of the siwalik and older tertiary rocks. The soil is mainly young loam rejuvenated every year by constant deposition of silt, clay and sand brought by different streams. This soil is deficient in phosphoric acid, nitrogen and humus, but potash and lime are usually present in sufficient quantity. There are three major types of soil in Bihar- (i) Piedmont swamp soil is found in northwestern part of west Champaran district, (ii) Terai soil, found in northern part of the state along the border of Nepal and (iii) The Gangetic alluvium in the plain of Bihar is covered by gangetic alluvium (both new as well as old). Six types of soils are present in Orissa. They are 'saline soils' found along some coastal strips; 'alluvial soils' in the coastal belt of Balasore, Cuttack, Puri and Ganjam districts as well as in the river valleys and flooded plains; 'laterite / lateritic soils' in the slopes of the hilly regions of Mayurbhanj down south to Koraput; 'red soils' in the north west of Cuttack district and Dhenkanal, Mayurbhanj, Keonjhar, Sundargarh and Bolangir districts; 'brown forest soils' in the districts of undivided Koraput and Kalahandi and 'black cotton soils' in Angul, Athmalik, Aska, Bolangir and Boudh. Soil resources of Jharkhand state mainly consist of in-situ soils formed from weathering of rocks and minerals. Soils are further divided into five groups, they are- (i) red soils found mostly in the Damodar valley and Rajmahal area, (ii) micacious soils (containing particles of mica) found in Koderma, Jhumeritilaiya, Barkagaon, and areas around the Mandar hill, (iii) sandy soils generally found in Hazaribagh and Dhanbad, (iv) Black soils, found in Rajmahal area and (v) laterite soils, found in western part of Ranchi, Palamau, and parts of Santhal Parganas and Singhbhum.

2.1.7 Vegetation

The fauna and flora of West Bengal possess the combined characteristics of the Himalayan and sub Himalayan Gangetic plain. Biodiversity is shaped by the biotic and abiotic components of its environment and this state has a rich assemblage of diverse habitats and vegetation designated with the help of eight different forest types. Diversity is further reflected in different types of ecosystem available here like mountain ecosystem of the north, forest ecosystem (semi-evergreen, deciduous, dry, moist and tidal varieties) extending over the major part of the state, freshwater ecosystem (rivers, streams, wetlands and to some extent estuaries), semiarid ecosystem in the western part, mangrove ecosystem in the south and coastal marine ecosystem along the shoreline. Orissa is facing environmental degradation worse than any other states in the country. The root cause of the disaster is the large-scale deforestation over the past few decades. Forests cover nearly 37 percent of the total area of the state. It is mainly divided into four categories viz. dense forest, sparse dense forest, tidal forest and fallow forest land. Bihar lies in the tropical to sub tropical region. Rainfall here is the most significant factor in determining the nature of vegetation. Bihar has a monsoon climate with an average annual rainfall of 1200 mm. The sub Himalayan foothill of Someshwar and Dun Rangesin Champaran constitute another belt of moist deciduous forests. These also consist of scrub, grass and reeds. Here the rainfall is above 1600 mm and thus promotes luxuriant Sal (*Shorea Robusta*) forests in the favored areas. The hot and dry summer gives the deciduous forests. The most important trees are Sal (*Shorea Robusta*), Shisham (*Dalbergia Sissoo*), Tun (*Cedrela Toona*), Khair (*Acacia Catechu*), and Simul (*Bombax ceiba*). This type of forests also occurs in Saharasa and Purnia districts. Jharkhand possesses rich floral heritage, a result of its being part of Chhotanagpur plateau. Main vegetation in this region comprises moist deciduous and dry deciduous forests. The vegetation includes Sal (*Shorea Robusta*) and bamboo (*Bambusa dendrocalmus*) as the major component.

2.2 Physiographic setting of the study area

India is a land of diversities. Great mountains, rivers, wide plateaus and plains, rolling uplands, lengthy coastlines etc., constitute the topography of our country. It has a monsoon climate with local and seasonal climatic diversities. Physiographically, India can be classified into four divisions (Figure 2.4), namely (i) the northern mountain region, (ii) the great plains of the north, (iii) the peninsular plateau and (iv) the coastal plains and islands (Jain et al. 2007).

The entire study area comes under the Indo-Gengetic plains, a major constituent of Indian subcontinent. The Indo-Gangetic plains, also known as the *Great Plains* are large floodplains of the Indus and the Ganga-Brahmaputra river systems. They run parallel to the Himalaya Mountains, from Jammu and Kashmir in the west to Assam in the east and draining most of northern and eastern India. The plains encompass an area of 700,000 sq km. The major rivers in this region are the Ganges and the Indus along with their tributaries– Beas, Yamuna, Gomti, Ravi, Chambal, Sutlej and Chenab. The great plains are sometimes classified into four divisions starting from the foot of mountains to the river mouth:

- ✓ The **Bhabar** belt is adjacent to the foothills of the Himalayas and consists of boulders and pebbles which have been carried down by the river streams. As the porosity of this belt is very high, the streams have substantial component of subsurface flow. The bhabar is generally narrow with its width varying from 7 to 15 km.

- ✓ The **Terai** belt lies next to the Bhabar region and is composed of newer alluvium. The underground streams reappear in this region. The region is excessively moist and thickly forested. It also receives heavy rainfall throughout the year and is populated with a variety of wildlife.
- ✓ The **Bangar** belt consists of older alluvium and forms the alluvial terrace of the flood plains. In the Gangetic plains, it has a low upland covered by laterite deposits.
- ✓ The **Khadar** belt lies in lowland areas after the Bangar belt. It is made up of fresh newer alluvium which is deposited by the rivers flowing down the plain.

The Indo-Gangetic belt is the world's most extensive expanse of uninterrupted alluvium formed by deposition of silt carried by the numerous rivers. The plains are flat making it conducive for irrigation through canals. The area is also rich in ground water sources. The plains are one of the world's most intensely farmed areas. The main crops grown are rice and wheat, which are grown in rotation. Other important crops grown in the region include maize, sugarcane and cotton. The Indo-Gangetic plains rank among the world's most densely populated areas.

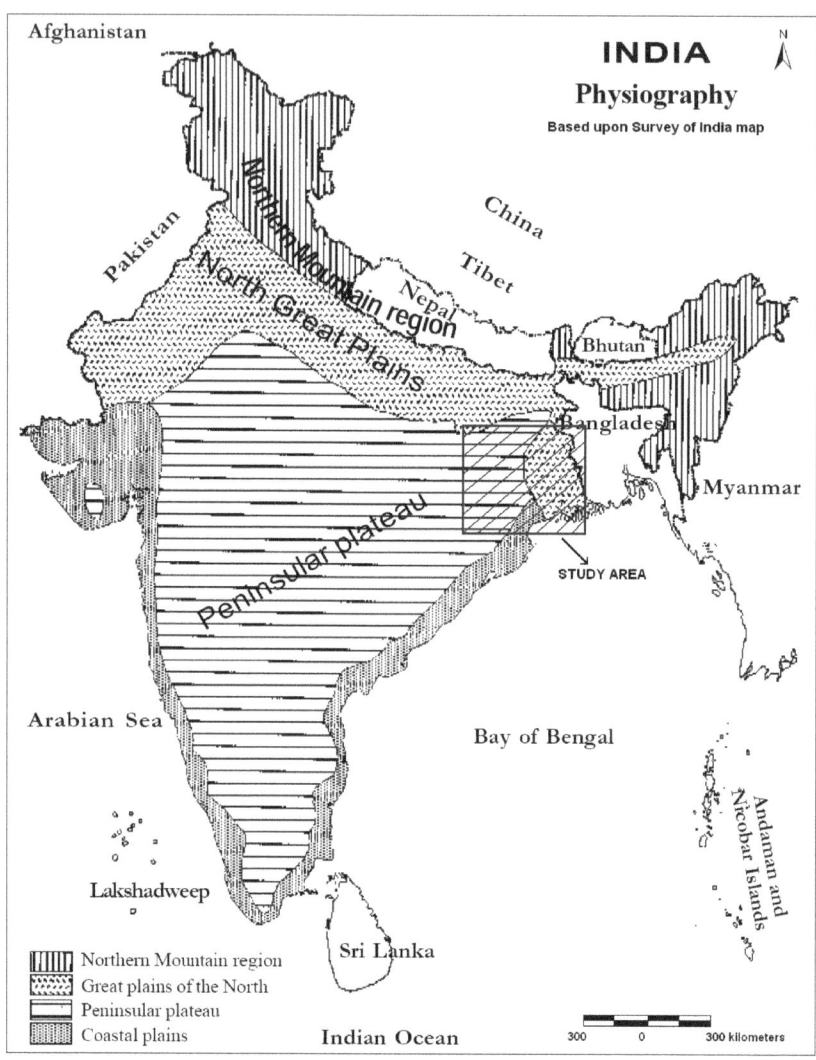

Figure 2.4. Physiographic division map of India

2.3 Review of literature

Although some great deal of work is already accomplished in the relevant field as mentioned in the underlying paragraphs, the study on climatological modelling in a GIS platform is indeed lacking. The thesis is a modest effort to fill in this gap. A large number of research works have been done on climate, climate change, climatological modelling in different region.

Jingyong Z et al. 2005 "Impact of Land Use Changes on Surface Warming in China"- this paper summaries the findings of Land use changes such as urbanization, agriculture, pasturing, deforestation, desertification and irrigation can change the land surface heat flux directly, and also change the atmospheric circulation indirectly, and therefore affect the local temperature. But it is difficult to separate their effects from climate trends such as greenhouse-gas effects. The results show that urbanization and other land use changes may contribute to the observed $0.12°$ C $(10 \text{ yr})^{-1}$ increase for daily mean surface temperature, and the $0.20°$ C $(10 \text{ yr})^{-1}$ and $0.03°$ C $(10 \text{ yr})^{-1}$ increases for the daily minimum and maximum surface temperatures, respectively. This study illustrates the possible impacts of land use changes on surface temperature over China.

Huqiang Z et al. 2006 "Impact of Land Use in China on regional climate: An Australia-China bilateral project on climate change"- this paper summaries land-surface processes and land-use form an essential component of the study of the Australia-Asian monsoon system and its potential changes in future climate. This study shows moderate but statistically significant regional impacts caused by land-use vegetation change. The surface climate cooling effect is primarily due to the increase in surface albedo, while the reduction in precipitation is associated with an enhanced southward penetration of dry and cold air due to reduced surface roughness.

Stohlgren T J et al. 1998 "Evidence that local land use practices influence regional climate, vegetation and stream pattern in adjacent natural areas"- in this paper author present evidence that land use practice in the plains of Colorado influence regional climate and vegetation in adjacent areas in the Rocky Mountains in predictable ways. Meso-scale climate model simulations using the Colorado State University Regional Atmospheric Modelling System projected that modifications to natural vegetation in the plain, primarily due to agriculture and urbanization, cloud produce lower summer temperature in the mountains. Result of the meso-scale climate model comparison between natural and current vegetation scenarios showed that current land use practice is the cause of cooler summer temperature.

Snyder M A et al. 2006 "Regional climate effects of irrigation and Urbanization in the Western United State: a model intercomparison"- The purpose of this study was to understand what effect conversion of natural vegetation to irrigated agriculture and urban areas has had on the climate of California and other areas of the western United States. The study compared climate effects produced by four regional climate models to identify consistent results. This study compares the climate responses of four regional climate models to past land-use changes in the Western United State. According to three regional climate models irrigation by supplementing soil moisture, producing large decreases in August mean and maximum temperatures where natural vegetation was converted to the irrigated agriculture. Conversion to irrigated agriculture also resulted in large increases in relative humidity (9% to 36%). Another regional climate model represents increases in summer minimum temperature due to converting natural vegetation to urban land cover.

Bounoua L et al. 1993 "Influence of Soil Moisture on the Sahelian Climate Prediction" – this summaries the influence of soil moisture on the local climate. An energy balance based statistical parameterization of the soil moisture availability has been developed and implemented in the Florida State University Global Spectral Model to test its performance in long range prediction. Specifically, a soil moisture parameter based on a moisture budget analysis has been introduced to estimate the Bowen ratio. A reduction of the surface roughness would not change the surface radiation. Likewise the fluxes of heat and moisture would remain almost the same as for a rough surface. However, since a decrease in the surface roughness results in a reduction of the bulk energy transfer coefficients, it must be accompanied by larger gradients of temperature and specific humidity between the surface and the lowest planetary boundary layer. The new soil moisture parameterization has also shown a realistic description of the West African monsoon surges over the Sahelian regions. When the monsoon moisture reached its northernmost position, a decrease in the net solar radiation reaching the ground led to a decrease in surface temperature.

Pan Z et al. 2001 "Soil moisture in a regional climate model"- this paper summaries how the soil moisture influence the regional climate. For several reasons soil moisture is one of the least known variables in climate simulation. Soil moisture is highly heterogeneous in space because of variability in soil type, landscape and precipitation. This variability severely reduces the representativeness of soil moisture measurement and complicates its parameterization in numerical models. Comparison of soil moisture for a run driven by a GCM 10-year enhanced greenhouse-gas scenario with that from a run driven by the corresponding control climate simulation indicates that soil becomes wetter in summer, probably because warm-season precipitation also increases in this scenario. In winter, the top 10 cm soil becomes drier while the top 1 m soil becomes slightly wetter. Errors in soil moisture correspond both in space and in time with precipitation errors, suggesting that much of the soil moisture error is attributable to a low bias of precipitation. The results indicate that improvement of soil moisture simulation will depend mainly on improvement in predicting precipitation, as well as improved representation of biophysical processes that control evapotranspiration.

Douglass D H et al. 2004 "Altitude Dependence of Atmospheric Temperature Trends"- This paper describe a positive temperature trend that is greater for the troposphere than the surface. This predicted positive trend increases in value with altitude until it reaches a maximum ratio with respect to the surface of as much as 1.5 to 2.0 at about 200-400 hecta Pascle (hPa). All three of the selected state-of-the-art climate model calculations show the same altitude dependence in all of the zonal plots for similar forcing scenarios. They give positive trend-lines at the earth's surface in all latitude bands. The values generally increase with altitude showing maximum trends at pressures of the order of 300 hPa. In particular, the state-of-the-art greenhouse models examined here show positive temperature trends that increase with altitude, reaching values greater than the near-surface trends by as much as 50 to 100 percent.

Pielkel R A et al. 1990 "Influence of landscape structure on local and regional climate"- This paper discusses the physical linkage between the surface and the atmosphere, and demonstrates how even slight changes in surface conditions can have a pronounced effect on weather and climate. Observational and modelling evidence are presented to demonstrate the influence of landscape type on the overlying atmospheric conditions. The albedo, and the fractional partitioning of atmospheric turbulent heat flux into sensible and latent fluxes is shown to be particularly important in directly affecting local and regional weather and climate. It is concluded that adequate assessment of global climate and climate change cannot be achieved unless mesoscale landscape characteristics and their changes over time can be accurately determined.

Klink K et al. 1994 "Influence of soil moisture and surface roughness heterogeneity on modeled climate"-This paper describe climatic importance of land surface variability, especially in soil moisture and surface roughness. Soil moisture influences the climate of a region directly through its affects on the latent heat flux. Evapotranspiration is enhanced when adequate soil water is available to plants. It also is true that the more radiative energy is used for latent heat flux, the less sensible heat is available within the near-surface environment. With an increase in depth of soil water (water table), evapotranspiration is reduced, and surface and air temperatures will increase. Surface roughness also plays an important role in the in surface energy balance by changing the efficiency of energy transport near the earth's surface.

Hartkamp A D et al. 1999 "Interpolation Techniques for Climate Variables"- This paper examines statistical approaches for interpolating climatic data over large regions, providing a brief introduction to interpolation techniques for climate variables of use in agricultural research, as well as general recommendations for future research to assess interpolation techniques. Three approaches- inverse distance weighted averaging; thin plate smoothing splines and co-kriging were evaluated for a 20,000 sq km square area covering the state of Jalisco, Mexico. Monthly mean data were generated for 200 meteorological stations and a digital elevation model (DEM) based on 1 sq km grid cells was used. Validation of the surfaces using two independent sets of test data showed no difference among the three techniques for predicting precipitation. For maximum temperature, splining performed best. Taking into account valued error prediction, data assumptions, and computational simplicity; authors recommend use of thin-plate smoothing splines for interpolating climate variables.

Bruse M and Skinner C J 1999 "Rooftop greening and local climate: a case study in Melbourne"-In this paper a high-resolution numerical model is used to estimate the microclimatic effect of introducing vegetation at street and/or rooftop level. When trees and soil give way to buildings and paved surfaces, the energy balance near the earth's surface changes. Less incoming solar radiation is dissipated as latent heat and more goes into sensible heat. The resulting rise in surface temperatures is one contributing factor to the urban heat island effect – an effect which is particularly unwelcome in Australia's rather warm climates. Results are presented to show the effect on a hot afternoon at a location in the inner suburbs of Melbourne, Australia. The study shows that the addition of vegetation reduces temperatures and wind speeds, thereby improving climatic amenity for pedestrians.

Weng Q 2001 "A remote sensing–GIS evaluation of urban expansion and its impact on surface temperature in the Zhujiang Delta, China"- The goal of this paper is to demonstrate the integrated use of remote sensing and GIS in addressing environmental issues in China at a local level and to evaluate urban growth patterns in the Zhujiang Delta and to analyze the impact of the urban growth on surface temperature. Urban development usually gives rise to a dramatic change of the Earth's surface, as natural vegetation is removed and replaced by non-evaporating and nontransparent surfaces such as metal, asphalt and concrete. This alteration will inevitably result in the redistribution of incoming solar radiation, and induce the urban–rural contrast in surface radiance and air temperature. Given the relationship between surface radiant temperature and the texture of land cover, the impact of urban development on surface temperature in the Zhujiang Delta can be assessed. The spatial pattern of radiant temperature increase was correlated with the pattern of urban expansion.

Tucker M R et al. 2001 "A comparison of Meteosat rainfall estimation techniques in Kenya"- this paper discuss how to estimate the rain fall through remote sensing satellite data over an area. Two methods for estimating ten-day rainfall totals from Meteosat infra-red imagery were compared for the April–June 1996 'long rains' of Kenya in an area covering the eastern highlands and the Tana and Athi river basins. One of these (the Bristol 'B4' method) was then used for rainfall estimation for the whole of Kenya, for November 1996 and the other, the 'Cold Cloud Duration' (CCD) method was used to estimate rainfall for the whole of Kenya for November 1997 to April 1998. For this comparison period the B4 method gave better estimates of actual rainfall than the CCD method because it used a variable cold cloud threshold temperature and ongoing calibration against rain gauge data. Meteosat images were converted for analysis in an image analysis package. The locations of Kenya synoptic meteorological stations were entered into the package. The pixel value of CCD was extracted automatically for each day of a month for each meteorological station. Output text values were converted into spreadsheets for analysis by comparison with rainfall station data.

Upmanis H and Chen D 1999 "Influence of geographical factors and meteorological variables on nocturnal urban-park temperature differences—a case study of summer 1995 in Goteborg, Sweden"- This paper presents a systematic statistical analysis, examining the magnitude of the influence of geographical factors and meteorological variables on the air temperature difference between a park and its built-up surroundings at nighttime in summer. Authors have successfully used principal component analysis to separate the variation in space and time, and have thereby been able to identify dominant spatial patterns and their associated evolution in time. They have established the different Geographical factors, such as surface properties and sky view factor (SVF) are primary reasons for the temperature difference, while meteorological variables, such as wind speed and cloud cover, are the forcing factors.

Menz G 1997 "Regionalization of precipitation models in East Africa using Meteosat data" - The article outlines the possibilities, prospects and limitations of estimating precipitation on the basis of high temporal resolution METEOSAT Real Time Window satellite data for the region of East Africa and the adjacent Indian Ocean. Authors have regionalized the study area for both the typical dry season and the rainy season using two distinct precipitation models. They have discussed natural conditions of topography, land-water ratio and the interrelation between highland and lowland areas within the study area as well as climatic/geographic factors of air-mass transport and cloud systems and have evaluated the effects of these variables to the precipitation.

Rigol J P, Jarvis C H and Stuart N 2001 "A neural network model for spatial interpolation of Rainfall data"- This paper examines the spatial and temporal distribution of rainfall from day to day and has a major impact on the agriculture, environment and economy of many countries. Spatially distributed estimates of rainfall are required inputs to many environmental models. Hydro-meteorological data such as air temperature, humidity and rainfall are typically recorded at the ground level at a limited number of weather stations scattered over a region. Consequently, values have to be estimated at intermediate un-sampled locations in order to generate continuous surfaces for a whole area. Author describe how the network was designed to accept a range of environmental data sets that are co-related to rainfall under certain conditions and the rainfall amounts observed at the nearest neighbor stations to each location. Author also describe that artificial neural networks (ANN) are particularly adept at handling massive amounts of data, dealing with complex nonlinear relationships, coping with non-normal and inter-correlated inputs, and allowing the incorporation of additional data and expert knowledge about a particular geographical domain within the estimation process.

Du Q et al. 2008 "Combination of multispectral remote sensing, variable rate technology and environmental modelling for citrus pest management"- this study demonstrates the potential of precision farming and agricultural sustainability. Precision agriculture is a production system that promotes variable management practices within a field according to site conditions, such as soil characteristics and weather conditions in order to adjust the inputs used and ultimately achieve optimal output. Precision farming incorporates several technological tools that include variable rate technology (VRT), remote sensing technologies, global positioning systems (GPS) and geographical information systems (GIS). Meteorological data such as temperatures, precipitation, evapotranspiration were used for the Pesticide Root Zone Model (PRZM) simulation. This study provides an example of how remotely sensed data and environmental models can be utilized in site-specific agricultural management for pest control and impact assessment. It quantitatively shows the advantage of using an integrated approach combining remote sensing, VRT, and environmental modelling for the characterization of environmental benefit in precision farming.

Servilla M et al. 2000 "Integrating High-Resolution Satellite Imagery and Weather Data for Improved Agricultural Management Decisions"- this paper analyzes high-resolution imagery with historic, current, and forecast weather data into a large-scale GIS map layer provides a powerful modelling and assessment tool that can lead to improved management decisions for agriculture. This study also demonstrates microclimate weather information can serve as a powerful tool to manage crops at the field and sub-field level. Imagery provides direct crop information by reference to location, while microclimate weather data provides both direct and indirect information about factors that can influence the health of crops.

CHAPTER 3

LAND USE /LAND COVER AND SOIL TEXTURE

CHAPTER OUTLINE

- Land use /land cover of the study area
- Soil texture of the study area
- Result and discussion

3.0 LAND USE /LAND COVER AND SOIL TEXTURE

3.1 Land use /land cover of the study area

Land cover is the natural landscape / anthropogenic supplement recorded as surface component such as forest, water, vegetation, wetlands, soils, rocks, and urban infrastructure (constructed materials covering the land surface) physically present and visible. As it is under view shed, it is easily amenable to remote sensing (Anderson et al. 1972). Thus land cover can be interpreted and documented by analyzing satellite and aerial imagery (Pal et al. 1992). This documentation of land cover based on spectral signatures is land cover mapping by dint of remote sensing (Anderson 1976). Land use can be defined as economic and cultural activities that are practiced at a place, but these activities may or may not be manifested as visible land cover features. Land use is the documentation of human uses of the landscape: residential, commercial, agricultural, etc. This documentation of land use can be inferred but not explicitly derived from satellite and aerial imagery. There is no spectral basis for land use determination in satellite imagery. For example, industrial or commercial land use may be visibly manifested as developed land cover, but recreational land use may occur in many different types of land cover, often without visible evidence of recreational use (Dobson et al. 1995).

3.1.1 Importance of land cover mapping

Identifying, delineating, and mapping land cover is important for resource management and planning programs. Land cover mapping is an essential tool used by natural resource managers as they struggle to protect habitat and plan against future loss and degradation. The challenge for resource managers is ever increasing as population growth in coastal regions takes its toll in the form of loss of wetlands and adjacent habitats, and as waste loads and competition for limited space and resources increase. Coastal wetlands are being destroyed by erosion, dredge, fill, impoundments and toxic pollutants (Dobson et al. 1995). Identification of land cover determines the baseline from which monitoring activities (land cover change detection) can be performed. Land cover change is a direct measure of quantitative habitat loss or gain (Dobson et al. 1995). Data portraying the characteristics and locations of land cover are necessary to ascertain which land areas and corresponding water resources are in critical need of analysis and protection. Land use and land cover change information can be used by resource managers in their decision-making processes to assess urban growth, determine changes to natural resources, and develop trend analyses. A variation of spatial air temperature is found between different land use/land cover categories on a diurnal basis and for all weather conditions (Eliasson et al. 2003).

3.1.2 Land use/ land cover classification categorization schemes

Broadly, land cover analyses can utilize field-based or remotely sensed data sources. Field-based land cover classification can extract detailed landscape information but is limited in its geographic scope. Time and labor constraints generally impact the extent of field data collection, and statistical analyses are used to determine the impact of sampling design on output significance. Remote sensing sources, such as satellite imagery or aerial photography, have a different set of output limitations, such as cost, spatial and spectral resolution, and interpretability (Jensen 2005a).

3.1.3 Land use/ land cover classification criteria

A land use and land cover classification system which can effectively employ orbital and high-altitude remote sensor data should meet the following criteria (Anderson 1971):

❖ The minimum level of interpretation accuracy in the identification of land use and land cover categories from remote sensor data should be at least 85 percent.

❖ The accuracy of interpretation for the several categories should be about equal.

❖ Repeatable or repetitive results should be obtainable from one interpreter to another and from one time of sensing to another.

❖ The classification system should be applicable over extensive areas.

❖ The categorization should permit vegetation and other types of land cover to be used as surrogates for activity.

❖ The classification system should be suitable for use with remote sensor data obtained at different times of the year.

❖ Effective use of subcategories that can be obtained from ground surveys or from the use of larger scale or enhanced remote sensor data should be possible.

❖ Aggregation of categories must be possible.

❖ Comparison with future land use data should be possible.

❖ Multiple uses of land should be recognized when possible.

Some of these criteria should apply to land use and land cover classification in general, but some of the criteria apply primarily to land use/land cover data interpreted from remote sensor data. Although land use data obtained at any level of categorization certainly should not be restricted to any particular level of user groups nor to any particular scale of presentation, information at Levels I and II would generally be of interest to users who desire data on a nationwide, interstate, or statewide basis. More detailed land use and land cover data such as those categorized at Levels III and IV usually will be used more frequently by those who need and generate local information at the intrastate, regional, county, or municipal level.

Level I land use information, for example, while efficiently and economically gathered over large areas by a LANDSAT type of satellite or from high-altitude imagery, could also be interpreted from conventional large-scale aircraft imagery or compiled by ground survey. This same information can be displayed at a wide variety of scales ranging from a standard topographic map scale, such as 1:24,000 or even larger, to the much smaller scale of the orbital imagery, such as 1:1,000,000.

Table 3.1. USGS level I land use/land cover classification system, description and color code

Level I (land use/land cover)	Description	color code
Urban or built-up land	Urban or built-up land is comprised of areas of intensive use with much of the land covered by structures. Included in this category are cities, towns, villages, strip developments along highways, transportation, power, and communications facilities	Red
Agricultural land	Agricultural land may be defined broadly as land used primarily for production of food and fiber.	Light brown
Rangeland	Rangeland historically has been defined as land where the potential natural vegetation is predominantly grasses, glasslike plants, forbs, or shrubs and where natural herbivore was an important influence in its pre-civilization state.	Light orange
Forest land	Forest lands have a tree-crown areal density (crown closure percentage) of 10 percent or more, are stocked with trees capable of producing timber or other wood products, and exert an influence on the climate or water regime.	Green
Water	Waters are those areas where the water is above the land surface for a significant part of most years.	Dark blue
Wetland	Wetlands are those areas where the water table is at, near, or above the land surface for a significant part of most years. The hydrologic regime is such that aquatic or hydrophytes vegetation usually is established.	Light blue
Barren Land	Barren Land is land of limited ability to support life and in which less than one-third of the area has vegetation or other cover. In general, it is an area of thin soil, sand, or rocks.	Gray
Tundra	Tundra is the term applied to the treeless regions beyond the limit of the boreal forest and above the altitudinal limit of trees in high mountain ranges.	Green-gray
Perennial Snow or Ice	Certain lands have a perennial cover of either snow or ice because of a combination of environmental factors which cause these features to survive the summer melting season.	White

3.1.4 Data uses for land use/ land cover data set preparation

Different types of data are used for land use/land covers map preparation, like- satellite imageries, topographical map and global land cover map of the study area.

(i) The LANDSAT program is a joint venture between the U.S. Geological Survey (USGS) and the National Aeronautics and Space Administration (NASA). The instrument on board LANDSAT 7 is the enhanced thematic mapper plus (ETM+). Enhanced instrument features in the LANDSAT 7 design allow monitoring of global, regional, as well as small-scale features and processes on the Earth's surface. Change detection studies for environmental, urban, or other applications are advanced by LANDSAT's range of spectral and spatial resolutions. Optical bands with standard false colour combination (SFCC) of LANDSAT 7 ETM+ satellite images are used to find out the land use/land cover classes in the study area.

(ii) University of Maryland Institute for advanced computer studies has developed global land cover classification product from satellite data for use in climate models. AVHRR data sets are resampled to a spatial resolution of one by one degree and used to carry out classification of global land cover. Classifications have also proceeded at a finer spatial resolution of 8 km at a continental scale. The data set contains 13 classes, namely- water, evergreen needle leaf forest, evergreen broadleaf forest, deciduous needle leaf forest, deciduous broadleaf forest, mixed forest, woodland, wooded grassland, closed shrub land, open shrub land, grassland, cropland, bare ground and urban and built-up. The data set is available in the band sequential format. All other details like-data spans along with the sources are given in the Table 3.2 and 3.3.

Table 3.2. Detail information of satellite images used for land use/land cover map preparation

Sl. No.	Satellite & sensor	Spectral bands	Path_ row	Date	Source
1			138 _43	14-10-2000	
2	LANDSAT-7	4th- NIR,	138 _44	02-11-2001	University of Maryland
3	ETM+	3rd - RED,	138 _45	02-11-2001	Institute for advanced
4		2nd -GREEN.	139 _43	26-10-2001	computer studies
5			139 _44	26-10-2001	
6			139 _45	08-11-2000	
7			140 _43	17-11-2000	
8			140 _44	17-11-2000	
9			140 _45	17-11-2000	

Table 3.3. Sources, year of publication and scale of topographical maps and other materials

Sl. No.	Topographical map and other materials	Scale	Year of publication	Source
1		1:250000	1960	University of Texas Libraries, Austin
2	Topographical map	1:50000	1973 - 1980	Survey of India
3	Land cover map	1:2000000	2001	University of Maryland Institute for Advanced Computer Studies

3.1.5 Identification of land use/land cover category

Keeping in mind, different land-use categories identifying on the image are classified inference made for the classification is as the follows:

Table 3.4. Land use/land cover category as visualized in LANDSAT-7 ETM+

Land use/land cover category	ETM+ Bands	Salient Characteristics
Clear water	7	Black tone in black and white and color.
Silty water	4, 7	Dark in 7; bluish in color.
Deciduous forests	5, 7	Very dark tone in 5, light in 7; dark red.
Coniferous forest	5, 7	Mottled medium to dark gray in 7, very dark in 5, and brownish-red and subdued tone in color.
Defoliated forest	5, 7	Lighter tone in 5, darker in 7, and grayish to brownish-red in color, relative to normal vegetation.
Mixed forest	4, 7	Combination of blotchy gray tones; mottled pinks, reds, and brownish-red.
Grasslands (in growth)	5, 7	Light tone in black and white; pinkish-red
Croplands and pasture	5, 7	Medium gray in 5, light in 7, and pinkish to moderate red in color depending on growth stage.
Moist ground	7	Irregular darker gray tones (broad); darker colors.
Soils-bare rock-fallow fields	4, 5, 7	Depends on surface composition and extent of vegetative cover. If barren or exposed, may be brighter in 4 and 5 than in 7. Red soils and red rock in shades of yellow; gray soil and rock dark bluish; rock outcrops associated with large landforms and structure.

Faults and fractures	5, 7	Linear, often discontinuous; interrupts topography; sometimes vegetated.
Sand and beaches	4, 5	Bright in all bands; white, bluish, to light buff.
Stripped land-pits and quarries	4, 5	Similar to beaches-usually not near large water bodies; often mottled, depending on reclamation.
Urban areas: commercial industrial	5, 7	Usually light-toned in 5, dark in 7; mottled bluish-gray with whitish and reddish specks.
Urban areas: residential	5, 7	Mottled gray, with street patterns visible; pinkish to reddish.

3.1.6 Methodology for preparation of land use/land cover map

The methodology to preparation of land use land cover data set of the study area is performed in four parts, as (i) pre-field study, (ii) laboratory work, (iii) field observation and verification and (iv) post-field laboratory work. Pre-field study includes- study of the background history of the study area which contains West Bengal, Orissa, Bihar and Jharkhand, study of the attributes contributing to the development of the existing physical environment, study of the land use land cover and collection of remote sensing and collateral data of the study area. Laboratory work includes geo-referencing, mosaicking, fusion, sub-setting, creation of masks taking the remote sensing and collateral data of the study area. Field observation includes ground truth collection, identification of different feature in different points and their spectral signature and post-field laboratory work includes digital classification, post classification verification, modification of the classification using ground truth, recode, filtering, accuracy and error matrix generation, statics generation according to estimated cell size for final data set in ASCII format and thematic map generation. Large area land use/land cover mapping using remotely sensed data needs careful planning of various activities. The following activities are particularly relevant to successful transformation of remotely sensed data into land use categories.

3.1.6.1 Collection of remote sensing data

At first remote sensing data are collected from sources primarily of data from the website of university of Maryland Institute for advanced computer studies. Selection of data depends on the level of land use/land cover mapping and the scale of the output map as visualized in the present study. Realizing the capability of remote sensing techniques a study was taken up to find out the land use/land cover classes in the study area using LANDSAT 7 ETM+ data of 2000-2001.

3.1.6.2 Collections and study of collateral data

Survey of India (SOI) topographical maps of 1:50000, topographic map of 1:250000 and other available information in the form of latest publications and maps pertaining to land use pattern serve useful reference material for planning ground truth collection. Some of the latest thematic and land use maps generated by NRSA, State remote sensing center, GSI and others agencies may serve useful reference material. In laboratory with the help of image processing S/W – ERDAS Imagine and ARC Map different tasks are performed, using satellite imageries.

3.1.6.3 Geo-referencing

Process of rectification involves geo-referencing that is assigning map co-ordinates to the satellite image. This is achieved by collecting ground control points from both the raw data (satellite) and the reference map (rectified already). The transformation process is carried out by estimating a suitable transformation relation between a set of points (GCPs) on the image as well as on a map (reference). When a specific map projection system is used to generate geo-coded image, the image needs to assign brightness values as well as pixel positions in the geo-coded map. Rectification of an image is always preceded by the determination of appropriate co-ordinate system for the database.

At first user performed single map rectification using the geographical Coordinate system and WGS 84 datum with a RMS error of 0.02. All the reference maps have been rectified by this process. Then the double image (Map to image/ image to image) rectification is performed for all the satellite imagery using the geographical Coordinate system and WGS 84 datum with a RMS error of 0.25. The process of rectification is performed in two steps:

    i.   Spatial interpolation (reassigning pixel positions)

   ii.   Intensity interpolation (reassigning pixel brightness value)

The errors induced in the image data due to attitude (pitch, roll and raw) or the altitude are best removed by identifying ground control points on the image as well as on the map and then mathematically modelling the geometric distortion present in the image.

Let the un-rectified image co-ordinate system is 'I, J', and the rectified co-ordinate system is (X, Y). The co-ordinate transformation may be modeled by a series of polynomial equations of appropriate order. Following is the example of a 2nd order polynomial, which is extensively used for geo referencing of raw images pertaining to varied terrain conditions. It is evident that most simple polynomial equation of 1st order can be solved using a minimum of 3 sets of GCPs which allows six equation (3 pairs) having 6 unknowns (constants) to solve. Here in this example of 2nd degree polynomial, there are '12' unknowns to solve. Thus we need to provide at least '6' GCPs to from 12 equations. But in practice user have to provide 'sample redundancies' in the form of at least '10 to 12' well distributed GCPs to include any co-ordinate extrapolations, which might give rise to untoward distortions in the resultant rectified scene.

$$I = a_0 + a_1 X + a_2 Y^2 + a_3 X^2 + a_4 XY + a_5 Y$$
$$J = b_0 + b_1 X + b_2 Y^2 + b_2 X^2 + b_4 XY + b_5 Y$$

The quadrilateral model necessitates a least square fit at nodes (corners), which results in some residual distortion at each point. The accuracy of transforms formation model is the order of polynomial and the relationship between the polynomial is given as –

$$N = (P + 1) \times (P + 2) \times 0.5$$

Where P is the order of the polynomial and N is the number of ground control points required.

After knowing the exact position (Co-ordinate) of the output pixels, we should find out the corresponding brightness value also i.e. we need to carry out the intensity interpolation. The desired position of the pixel of input scene does not fall exactly in the output grid location; the intensity of the pixel in the output can be interpolated from the input scene. There are 3 methods of interpolation techniques-

a. Nearest neighbor $\rightarrow$ in this method nearest pixel is selected of the map (pixel value remains unaltered only in this method).

b. Bilinear $\rightarrow$ Taking 4 pixel surrounding the I J co-ordinate with orthogonal distance (weighted average is taken for computation).

c. Cubic convolution → Taking equation distance of 16 surrounding pixel of I, J co-ordinate (weighted average is taken for computation).

Only the nearest neighborhood method as intensity interpolation is preferred to bilinear and cubic convolution when the pixel value changes in the latter methods are unsuitable for geo-referencing of classified output. In case of classified images, nearest neighbor method is invariably chosen as the method of intensity interpolation.

3.1.6.4 Mosaicking and sub-setting

The entire study area does not appear in a single satellite scene. All the geo-referenced images of different path/row are mosaicked to get the entire study area. Using the similar projection as well as datum for all the images, the mosaicking tool does the mosaicking effortlessly. The study area (21° N - 25° N and 85° E - 89° E) is extracted by sub-setting from the mosaic image using the co-ordinates of four corners.

3.1.6.5 Creation of AOI layer and masks

To extract the interested area for better classification masking has performed using AOI layers by the help of AOI tools in the ERDAS IMAGINE S/W on the satellite imagery.

3.1.6.6 Ground truth (GT) collection

A lot of homework is pre-requisite for planning a ground truth visit. The ground information is intended to serve, both as an aid for classification and as a reference to assess the accuracy of classifications and validation of the output. Precise location of training fields or GT blocks and proper timing, those are very important for GT collection. The ideal date, for acquisition of remotely sensed data and ground truth, should be compatible with vegetative growth of crops. The broad knowledge of phonologic behavior of the major crops (crop calendar) of the area is helpful in planning for GT collection. The ground truth blocks should be scattered over the entire lengths and breadth of the area to be mapped. It should cover all the different variations of soils and landscapes as the nature of soils and the relief features greatly influence the spectral signature of the various land use categories. A more practical and better alternative for this purpose is to use standard FCC and mark the tentative GT blocks covering all the distinct signatures on the FCC itself. The sites are to be transferred to the SOI toposheet as well, in order to enable the GT collecting scientist to reach the block in the field and collect the required ground information. Total hundred (100) land use land cover samples are identified with their associate feature in the field (Appendix 1). About 25% of the GT brocks are used for training sets and the rest 75% are used for assessment of the accuracy results (Anon 1991).

3.1.6.7 Digital image processing

Commonly used procedure is one supervised technique using maximum likelihood algorithm. Here the ground truth information is fed the computer and is called training process. Logic- is that the computer is trained with the different signatures for different categories (Jensen 1996). The second step is to generate statistics of individual categories like mean, standard deviation and variance-covariance matrix. Lastly the computer uses the class statistics to classify the entire scene (area) into various land use / land cover categories.

3.1.6.8 Classification

Information categories are generated from processing remote sensing data is called image classification. Taking the mosaicked and rectified subset images supervised classification is performed. From the tonal values of original FCC imageries, land cover classes are selected and for classification a signature editor is created to supervise the computer S/w. This classification is performed taking six classes. These are as the followings:

      i.   Water (Sea or inland water bodies)

      ii.   Marshy land

      iii.   Forest

      iv.   Shrub land (Mixed or open)

      v.   Crop land (Agricultural land)

      vi.   Built up area (Settlement or industries)

3.1.6.9 Post classification verification

It is important to evaluate the accuracy of classification. This is normally done by a reconnaissance trip in the terrain classified and by some random checking in different sites. Also it can be checked in one ground visit as explained in GT correction.

3.1.6.10 Recode

After coming back from field area with final ground truth info, different classification errors are attended to and are set right as post-classification correction. Then recode and accuracy assessment are performed. In geographic information system analyses recode is a process, which allows assigning a new class value number to one or many classes of an existing image file for creating a new output file. This function can also be used to combine classes by recording more than one class to the same new class number. With the help of ERDAS IMAGINE s/w recode is performed to the classified product of 2000-2001. After recoding each and every class has got a new set of class number, as Table 3.5.

Table 3.5. Recode values for different land use/ land cover classes

Row No.	Land use/ land cover name	Recode Value
1	Water (Sea or inland water bodies)	0
2	Marshy land	1
3	Forest	2
4	Shrub land (Mixed or open)	3
5	Crop land (Agricultural land)	4
6	Built up area (Settlement or industries)	5

3.1.6.11 Accuracy assessment

Classification accuracy assessment is performed in order to make the remote sensing derived land use or land cover map and associated statistics useful. Accuracy assessment is often an afterthought rather than an integral part of many remote-sensing studies. Non-site-specific accuracy assessments completely ignore locational accuracy. In other word, only the total amount of a category is considered without regard for the veracity of its location. A non-site-specific accuracy assessment often yields very high accuracy but misleading results when all the errors balance out in a region (Jensen 2005b). To correctly perform classification accuracy assessment, it is necessary to compare two sources of information: the remote sensing derived classification map, and the reference test information or ground truth. With the help of ERDAS s/w accuracy assessment is performed using the ground truth on the classified image (Appendix 2.1 and 2.2). In stratified sampling, the total number of pixels is subdivided into several classes, called strata, and then a sub samples combined together give the stratified sample. If the selection from strata is done by random sampling, the method is known as stratified random sampling. The subdivisions of total pixels into strata are done by purposive method, but the selection of sub-samples from within the strata depends purely on chance. Stratified random sampling may therefore be viewed as a mixture of both purposive and random sampling and combines the both. Stratified sampling is generally used when the total pixel is heterogeneous but can be subdivided into strata within each of which the heterogeneity is not prominent.

The Kappa coefficient measures the relationship between beyond chance agreement and expected disagreement. This measure uses all elements in the matrix and not just the diagonal ones (Lillesand et al. 2000). The estimate of Kappa is the proportion of agreement after chance agreement is removed from consideration: $K^\wedge = (po - pc) / (1 - pc)$

> po = Proportion of units which agree, =spij = overall accuracy.
> pc = Proportion of units for expected chance agreement = spi + p + i.
> pij = eij / NT.
> pi+ = row subtotal of pij for row i.
> p+i = column subtotal of pij for column j.

One advantage of using this method, that we can statistically compare two classification products. For example, two classified maps can be made use of by different algorithm and we can use the same reference data to verify them.

3.1.6.12 Classification errors matrix

The relationship between these two sets of information is commonly summarized in an error matrix. This is the most common means of expressing accuracy. It is prepared to determine how a classification has categories representative subject of pixel used in the training process of supervised classification. The classification error matrix shows the accuracy level of each classified image of different years (Lillesand et al. 2000). Omission errors correspond to non-diagonal column element. If some pixels of a specific class become omitted in correspond classified class then this type of error comes. Like: some pixels that should have been classified as "vegetation" are omitted from that category. Commission errors are representing by non-diagonal row elements. If some pixels of a specific class become included in other class then this type of error comes. Like: some pixels that should have been classified as "vegetation" are misclassified and added to the "settlement".

Over all classification accuracy is computed by dividing the total number of correctly classified pixel (diagonal elements) by the total number of reference pixels. Accuracy can be calculated for each category individually by dividing the number of correctly classified pixels in each category by the total number of reference pixels in the corresponding class (the column total) called producer's accuracy. User's accuracy can be calculated by dividing the number of correctly classified pixel in each category by the number of pixels that are classified in that category (the row total).

3.1.6.13 Generation of output

The classified output was subjected to smoothening using median filter. 3 x 3 kernels are used in the median filtering in stages to get the smoothened output. After obtaining satisfied level of accuracy, we have proceeded with the generation of the final output. Three sets of data/map are generated to represent the land use/land cover, as first one with 30m, second one with 1km and third one with 5km spatial resolution. Any output with profound visual contrast is always desirable. Therefore, a contrasting color look up' table (LUT) is developed in the computer for each class by assigning desired RGB against the code (Table 3.6). Land use land cover map is thereby generated (Figure 3.1, 3.2 and 3.3) wherein individual categories get distinct color.

Table 3.6. RGB color presentations and statistics for different land use/ land cover classes

Recode Value	Land use/ land cover	RGB Color presentation	Area (in sq km)	% of area
0	Water (Sea or inland)	0.000:0.000:1.000	19782	10.93
1	Marshy land	0.250:0.880: 0.820	2818	1.56
2	Forest	0.000:0.390:0.000	23048	12.73
3	Shrub land (Mixed or open)	1.000:0.750:0.800	22078	12.19
4	Crop land (Agricultural land)	1.000:1.000:0.120	112348	62.04
5	Built up area (Settlement or industries)	0.440:0.040:0.200	989	0.55

As can be seen in Table 3.6, 12.73 percent of the study area is covered by forest. For reference purposes, forest land cover is shown on the land use/land cover map as green in color. Forest cover is most heavily concentrated in the western part of the study area. About 12 percent land of the study area is covered by shrub, shown in pink color in the map. Agriculture activity is very important in the study area, almost in 62 percent land. It is shown as yellow in the land use/land cover map. Less than 1 percent land is used for residential and industries purpose, shown as brown color. Other land use/land cover, like inland and sea water is almost 11 percent and marshy land less than 2 percent are found in the area, shown as blue and turquoise color (Figure 3.1, 3.2 and 3.3). Water class is most heavily concentrated in the southern part and marshy land in the eastern part of the study area.

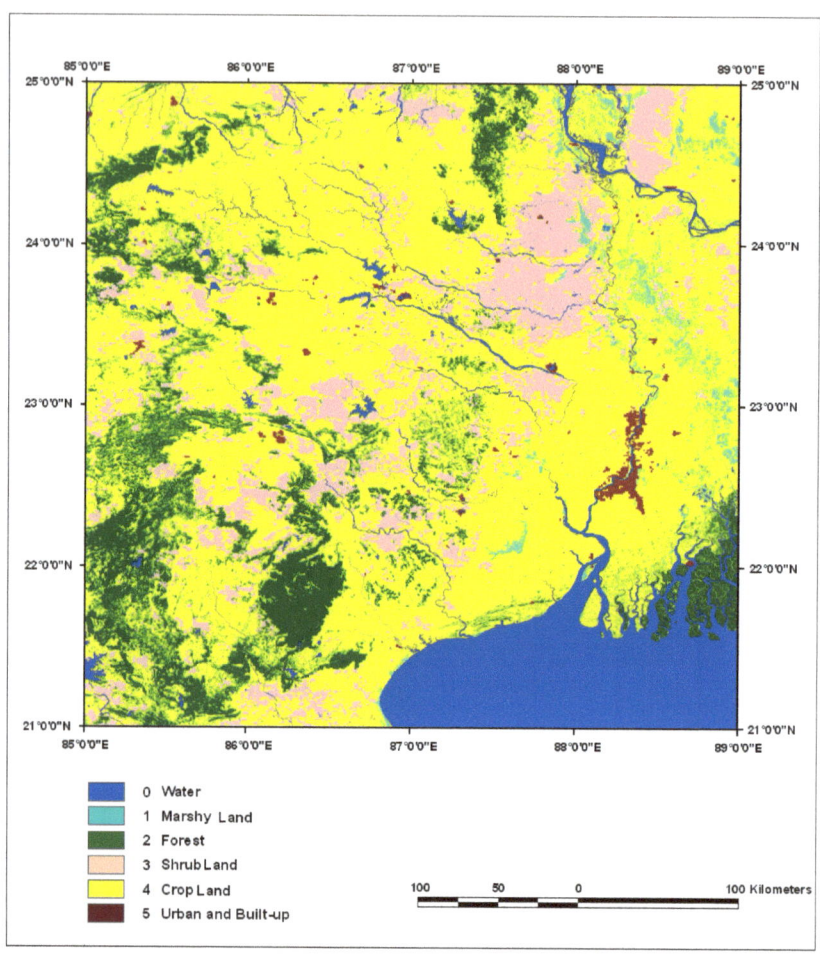

Figure 3.1. Land use/land cover (30 m x 30 m cell size) of the study area, generated from
LANDSAT-7, ETM+ satellite image

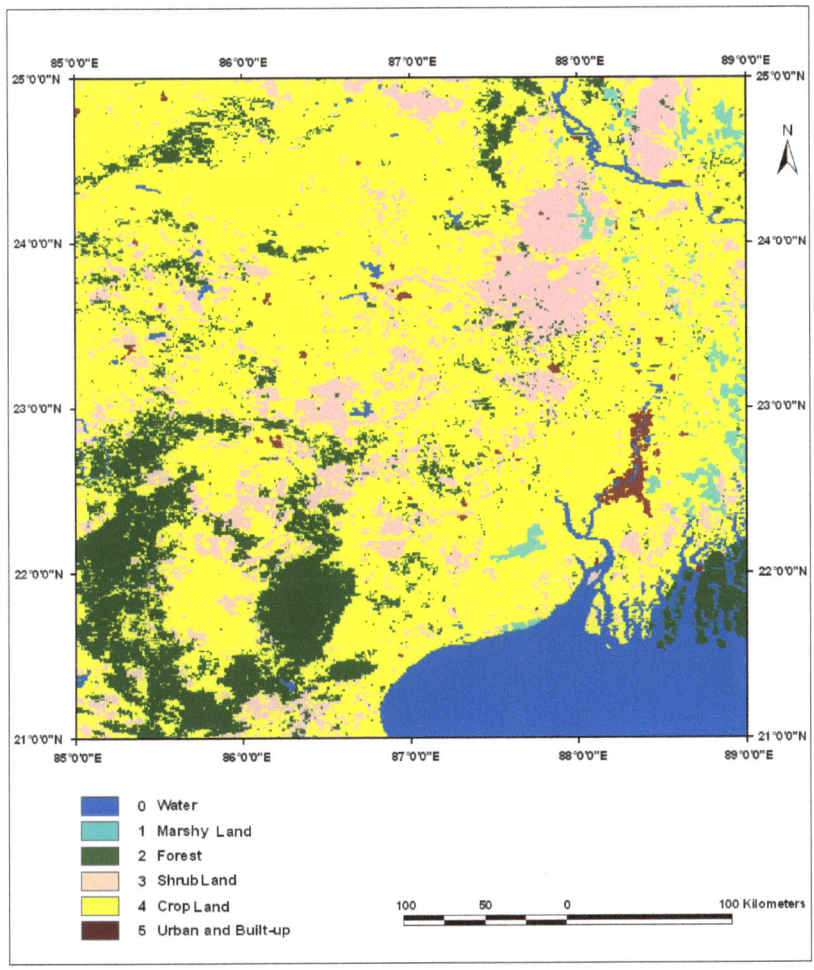

Figure 3.2. Land use/land cover (1 km x 1 km cell size) of the study area, generated from LANDSAT-7, ETM+ satellite image

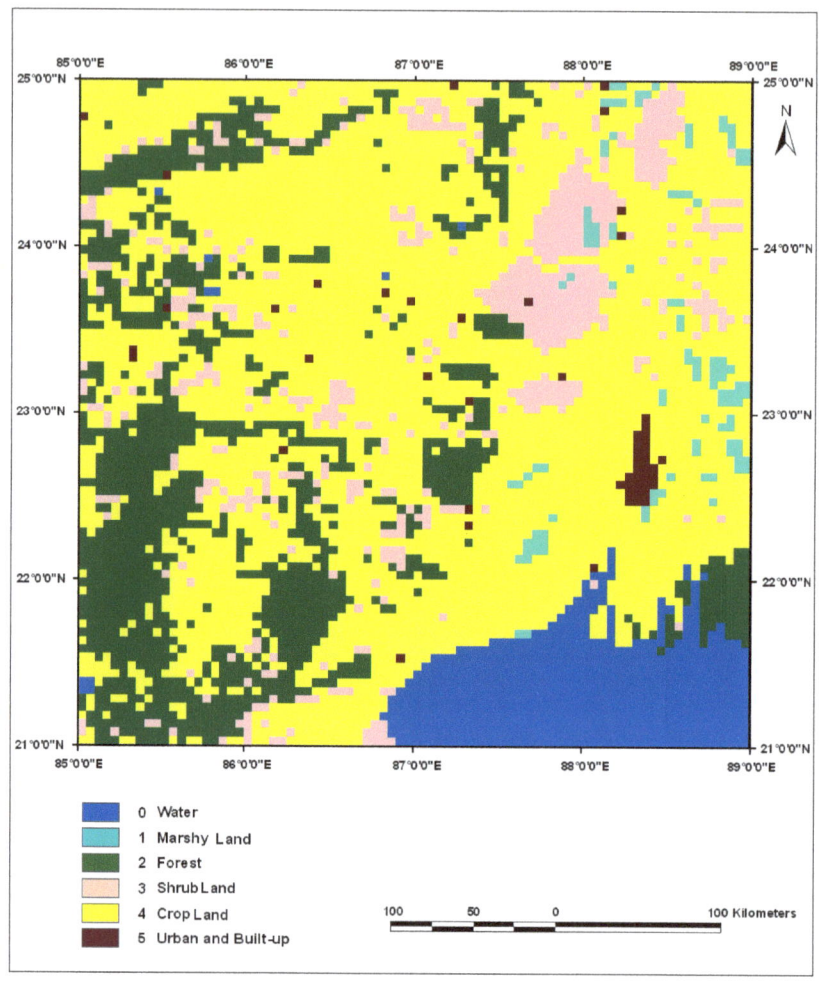

Figure 3.3. Land use/land cover (5 km x 5 km cell size) of the study area, generated from LANDSAT-7, ETM+ satellite image

3.1.7 Methodological flow chat for preparation of land use/land cover map

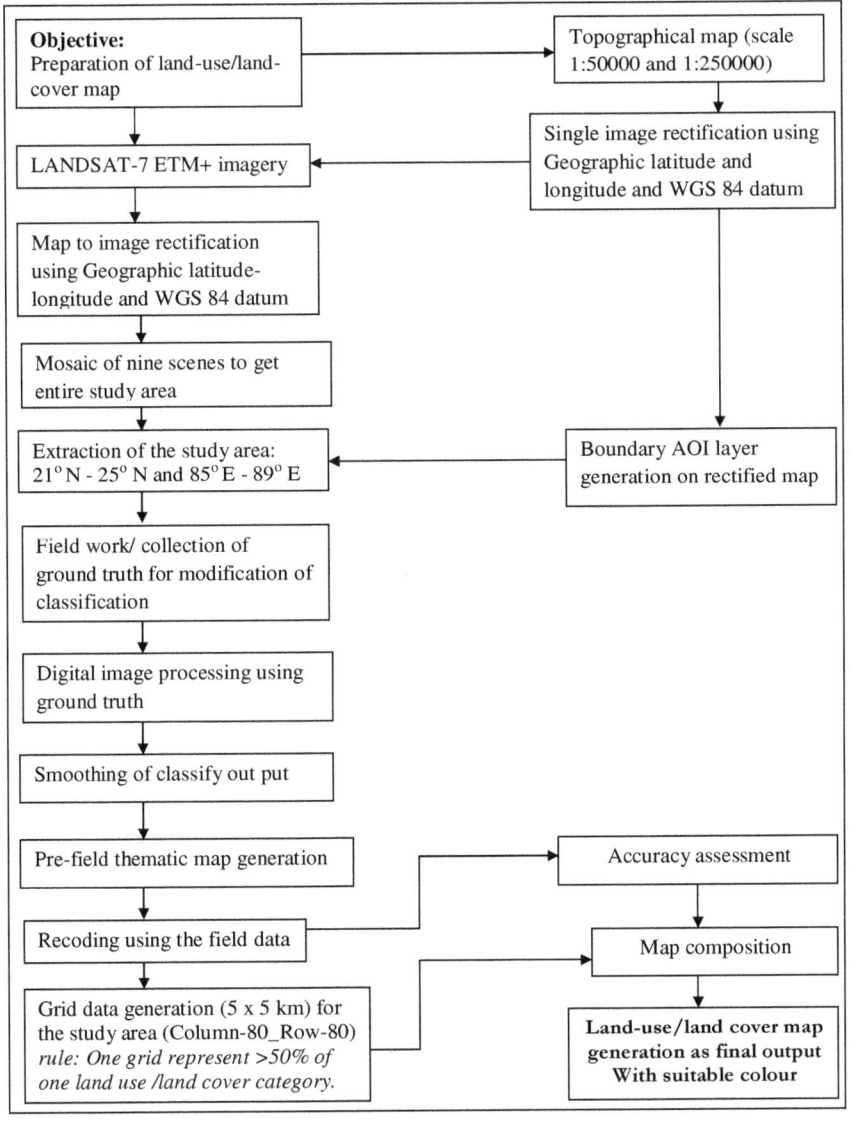

3.2 Soil texture of the study area

Soil happens to be the most important factor influencing the productivity of our lithosphere. Soil itself is very complex. Soil also contains air, water, dead organic matter, and various types of living organisms. The formation of a soil is influenced by organisms, climate, topography, parent material, and time. Four basic components that the soils are made of are mineral particles, water, air, and organic matter. Organic matter can be further sub-divided into humus, roots, and living organisms (Pidwirny 2006).

3.2.1 Soil texture

The texture of a soil refers to the size distribution of the mineral particles found in a representative sample of soil. Particles are normally grouped into three main soil separates: sand, silt, and clay in tune with diminishing size. Soil texture depends on the relative proportion of the three classes of soil separates in the soil. Sand is the largest particle and they feel "gritty". Silt is medium sized, and they feel soft, silky or "floury". Clay is the smallest sized particles, and they feel "sticky" and cohesive and they are hard to squeeze when dry. Table 3.7 describes the classification of soil separates according to size.

Table 3.7. Particle size ranges for sand, silt, and clay according to USDA

Type of mineral particle	Size range
Sand	2.0 - 0.05 millimeters
Silt	0.05 - 0.002 millimeters
Clay	less than 0.002 millimeters

3.2.2 Identification of soil type

Soils are generally described according to the predominant type of soil particle present - sand, silt, or clay. By conducting a simple soil test, we can easily see what kind of soil we're dealing with. We can perform repeatedly this test with several different soil samples from different location of the study area as the following:

- Filling up a quart jars about 1/3 full with topsoil and rest part by water.
- Screwing the lid and shake the mixture vigorously, until all the clumps of soil have dissolved.
- Now setting up the jar on a window sill and watch as the larger particles begin to sink to the bottom.

- In a minute or two the sand portion of the soil will have settled to the bottom of the jar.
- After several hours the finer silt particles will gradually settle onto the sand. Now we can find out the layers are slightly different colors, indicating various types of particles.
- The next layer above the silt will be clay.

3.2.3 Soil texture classification

Soil textures are classified by the fractions of each soil separate (sand, silt, and clay) present in a soil. Classifications are typically named for the primary constituent particle size or a combination of the most abundant particles sizes, e.g. "sandy clay" or "silty clay." A fourth term, loam, is used to describe a roughly equal concentration of sand, silt, and clay, and lends to the naming of even more classifications, e.g. "clay loam" or "silt loam." In the United States, twelve soil texture classifications are defined by the USDA: sand, loamy sand, sandy loam, silt loam, loam, sandy clay loam, silty clay loam, clay loam, clay, sandy clay, silty clay and Silt. Determining the soil textures is often aided with the use of a soil texture triangle. The USDA soil texture triangle is shown in the Figure 3.4.

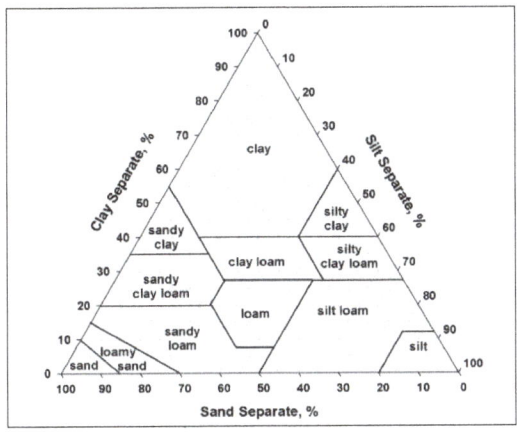

Figure 3.4. USDA Soil texture triangle

3.2.4 Data uses for soil texture map preparation

West Bengal soils sheet of National Bureau of Soil Survey & Land Use Planning (NBSS & LUP), soil region map of National Atlas & Thematic Mapping Organization (NATMO) with different group of soil and topographical map of the corresponding area are used to generate soil texture map. (i) NBSS & LUP has published the soil map on 1:500000 scale covering the entire state of West Bengal in four sheets. This data set provides brief description highlighting soil depth, texture, drainage, slope, erosion, salinity etc. of the dominant and associated subdominant mapped soil. The map also provides taxonomy of soil as per USDA system of classification for national and international understanding. (ii) NATMO has prepared national soil atlas map of India on 1:2000000 scale. This map provides information on 8 different groups (and 30 subgroups) of soil for eastern part of India, namely-entisols, vertisols, inceptisols, aridisols, mollisols, alfisols, ultisols and histosols orders. All other details of the data along with the sources are given in the Table 3.8.

Table 3.8. Sources, year of publication and scale of soil maps and topographical maps

Sl. No.	Name of materials	Scale	Year of publication	Source
1	West Bengal soil sheet (1 to 4)	1:500000	1981	National bureau of soil survey and land use planning
2	National soil atlas of India	1:2000000	1981	National atlas & thematic mapping organisation, department of science and technology, Govt. of India, Kolkata
3	Topographical	1:250000	1960	University of Texas libraries, Austin
4	map	1:50000	1973 - 1980	Survey of India, Kolkata

3.2.5 Methodology for preparation of soil texture map

The methodology to preparation of soil texture data set of the study area is performed in four parts- as I-pre-field study; II-laboratory work; III-field observation; and IV-post-field laboratory work. Pre-field study includes- study of the soil characteristics and collection of soil maps and collateral data of the study area. Laboratory work includes geo-referencing, mosaicking, fusion, sub-setting, creation of masks taking the soil maps and collateral data of the study area. Field observation includes ground truth collection, identification of different feature in different points and soil sample collection. Post-field laboratory work includes soil sample testing, classification of soil into different texture classes according to different characteristics, modification of the classification using ground truth, recode, filtering, accuracy and error matrix generation, statics generation according to estimated cell size for final data set in ASCII format and thematic map generation.

3.2.5.1 Collections of primary and collateral data

At first soil maps are collected as sources of primarily data from NBSS & LUP and NATMO, department of science and technology, govt. of India, Kolkata. Selection of data depends on the level of soil texture mapping and the scale of the output map as visualized in the present study. Survey of India (SOI) topographical maps of 1:50000, topographic map of 1:250000 and other available information in the form of latest publications serve useful reference material for planning ground truth collection. Some of the latest thematic and soil maps, generated by NBSS, NATMO, GSI and others agencies serves as useful reference material. In laboratory with the help of ArcGIS s/w different tasks are performed, using the available soil maps.

3.2.5.2 Geo-referencing, mosaicking and sub-setting

At first single map rectification is performed using the geographical co-ordinate system and WGS 84 datum with a RMS error of 0.02. ERDAS IMAGINE 9.0 is used to rectify topographical maps of the study area. All other reference maps have been rectified by this process. Then the double image (Map to map) rectification is performed for all soil maps using the geographical co-ordinate system and WGS 84 datum with a RMS error of 0.25. The entire study area appears in multiple sheets of soil map, hence is the need for mosaicking. All the geo-referenced map sheets of the area are mosaicked to get the entire study area. Using the similar projection as well as datum, all the soil maps are mosaicked in the ERDAS IMAGINE 9.0 software environment. The study area (21° N - 25° N and 85° E - 89° E) is extracted by sub-setting from the mosaic map using the coordinates of four corners.

3.2.5.3 Ground truth (GT) collection

The ground information is intended to serve, both as an aid for classification and as a reference to assess the accuracy of classifications and validation of the output. Precise location of training fields or GT blocks is very important. Equally important is the proper timing of the GT collection. The ground truth blocks should be scattered over the entire lengths and breadth of the area to be mapped. It should cover all the different variations of soils and landscapes as the nature of soils and the relief features greatly influence the spectral signature of the various land use categories. Total fourty nine (49) soil samples are collected to find out the soil texture of the specific place. After coming from field all the collected soil sample are tested to find out the percent of sand, silt and clay. According to the percent of sand, silt and clay (Table 3.9) all the soil samples are classified into different texture classes (Appendix 3).

Table 3.9. USDA soil texture classes description

Sl. No.	Texture classes	Description (USDA)
1	Sands	More than 85% sand, 0 to 10% clay and 0 to 15% silt
2	Loamy Sands	70 to 91% sand, 0 to 15% clay and 0 to 30% silt
3	Sandy Loams	More than 43% sand, less than 7% clay, less than 50% silt
4	Sandy Clay Loam	More than 45% sand, 20 to 35% clay, less than 28% silt
5	Sandy Clay	45% or more sand and 35% or more clay
6	Loam	Less than 52% sand, 7 to 27% clay, 28 to 50% silt
7	Silt Loam	Less than 12% clay and 50 to 80% silt
8	Clay Loam	20 to 45% sand and 27 to 40% clay
9	Silt	Less than 12% clay and 80% or more silt
10	Silty Clay Loam	Less than 20% sand and 27 to 40% clay
11	Silty Clay	40% or more clay and 40% or more silt
12	Clay	Less than 45% sand, 40% or more clay and less than 40% silt

3.2.5.4 Soil texture classification from soil type and soil region maps of WB and India

Soil texture classification is performed using two referenced soil maps, West Bengal soils sheets 1 to 4 of NBSS & LUP and national atlas of India, soil region of NATMO, department of science and technology, Govt. of India, Kolkata. All the soils are categorized into nine (9) soil texture classes according to their characteristics (Appendix 4).

3.2.5.5 Vector layer generation, digitization and attribution

To generate the soil texture map of the study area, ArcInfo and ERDAS IMAGINE s/w are used. First, create all vector layers (arc coverage) of soil texture category by digitization process in the ArcInfo and ERDAS IMAGIE s/w environment. With the help of poly-line tools different soil texture area are digitized. Then using 'CLEAN' followed by 'BUILD' command the poly-line layer is converted into polygon coverage. According to the soil texture classification from soil type and soil region maps of WB and India all the data are introduced into the attribute table of the polygon layer against each polygon. Another point layer is generated in the same s/w environment and the collected soil sample points are plotted according to the longitudinal and latitudinal value. Then soil texture attribute for each sample location is introduced to the point layer.

3.2.5.6 Vector to raster conversion

Using ERDAS IMAGINE s/w, the polygon coverage is transferred into a raster layer by vector utility dialog. The column with soil texture data is used to convert the soil texture layer from vector to raster format. All point data are interpolated to the raster through krigging interpolation techniques in the ArcGIS s/w environment.

3.2.5.7 Recode, accuracy assessment and output generation

After coming back from field with ground truth info, different classification errors are indentified and set right. Next recoding and accuracy assessment are performed. In geographic information system analyses recode is a process, which allows assigning a new class value to one or many classes of an existing image file for creating a new output file. This function can also be used to combine classes by recoding more than one class to the same new class number. With the help of ERDAS IMAGIE s/w recode is performed to the soil texture product of the area. After recoding, each and every class gets a new set of class number (Table 3.10). Accuracy assessment is performed using the ground truth (Appendix 3) on the recode soil texture data with the help of ERDAS IMAGIE s/w. After obtaining satisfactory level of accuracy, we proceeded with the generation of the final output. Two sets of data/map are generated to represent the soil texture, as first one with 500 m (Figure 3.5) and second one with 5 km spatial resolution (Figure 3.6). A contrasting color look up' table (LUT) is developed in the computer for each class by assigning desired RGB against the code (Table 3.10). A soil texture map is thereby generated wherein individual categories get distinct color.

Table 3.10. Recode values, RGB color presentations and statistics for soil texture classes

Row No	Soil texture classes name	Recode Value	RGB Color presentation	Area (in sq km)	% of area
1	Water	0	0.000:0.000:1.000	16422	9.1
2	Loamy sand	1	1.000: 1.000: 0.880	11467	6.3
3	Sandy loam	2	0.930:0.510: 0.930	37742	20.8

4	Silty loam	3	1.000:0.840:0.000	1302	0.7
5	Loamy	4	0.750:0.750:0.750	4360	2.4
6	Sandy clay loam	5	0.690:0.190:0.380	54248	30.0
7	Silty clay loam	6	0.650:0.160:0.160	27662	15.3
8	Clay loam	7	0.500:1.000: 0.000	1557	0.9
9	Sandy clay	8	0.000:0.390:0.000	5238	2.9
10	Silty clay	9	1.000:1.000:0.000	21065	11.6

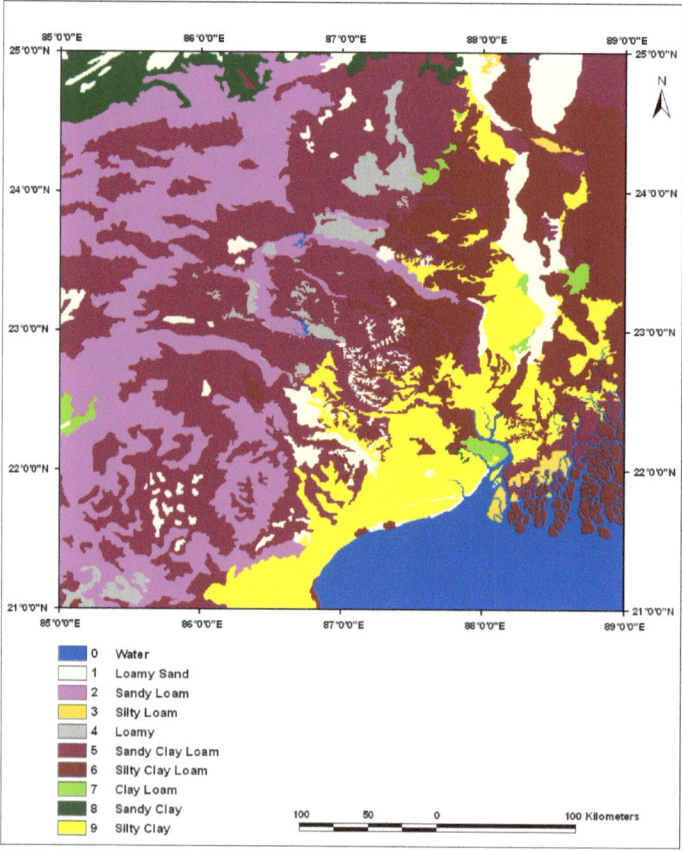

Figure 3.5. Soil texture map of the study area (0.5 km x 0.5 km) generated from soil region map and NBSS soil sheet

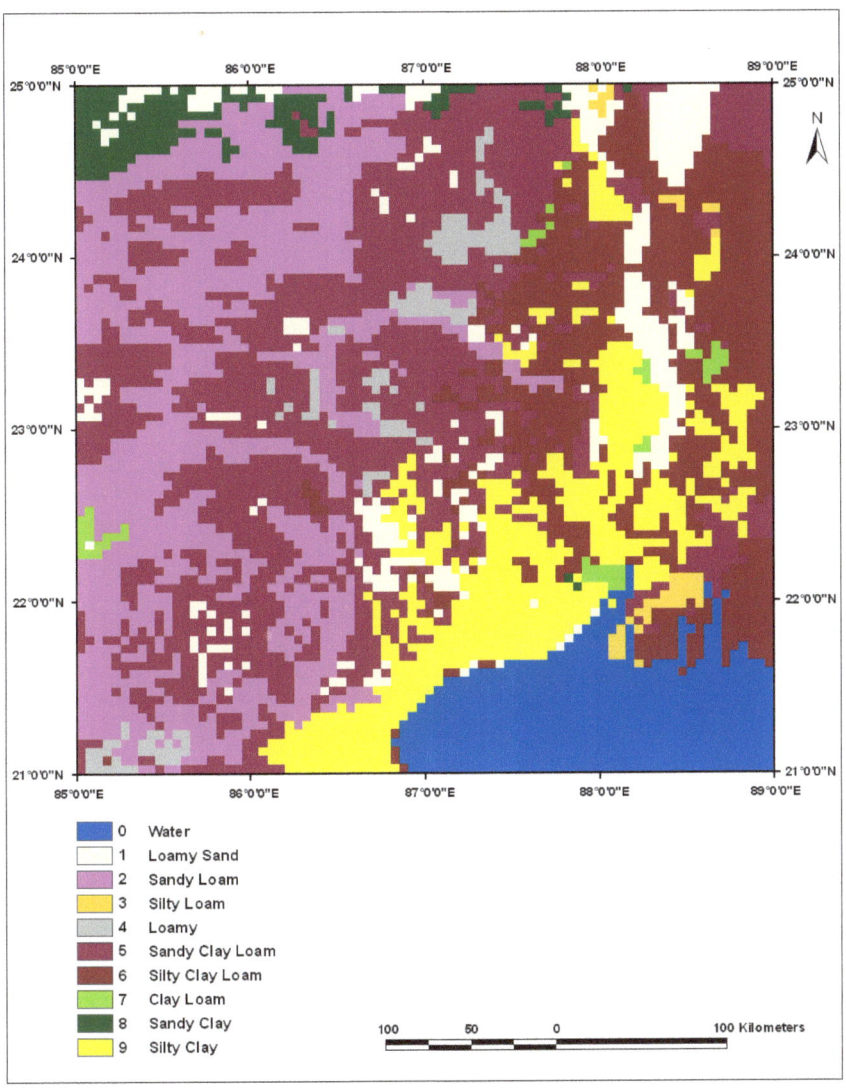

Figure 3.6. Soil texture map of the study area (5 km x 5 km) generated from soil region map and NBSS soil sheet

3.2.6 Methodological flow chart for preparation of soil texture map

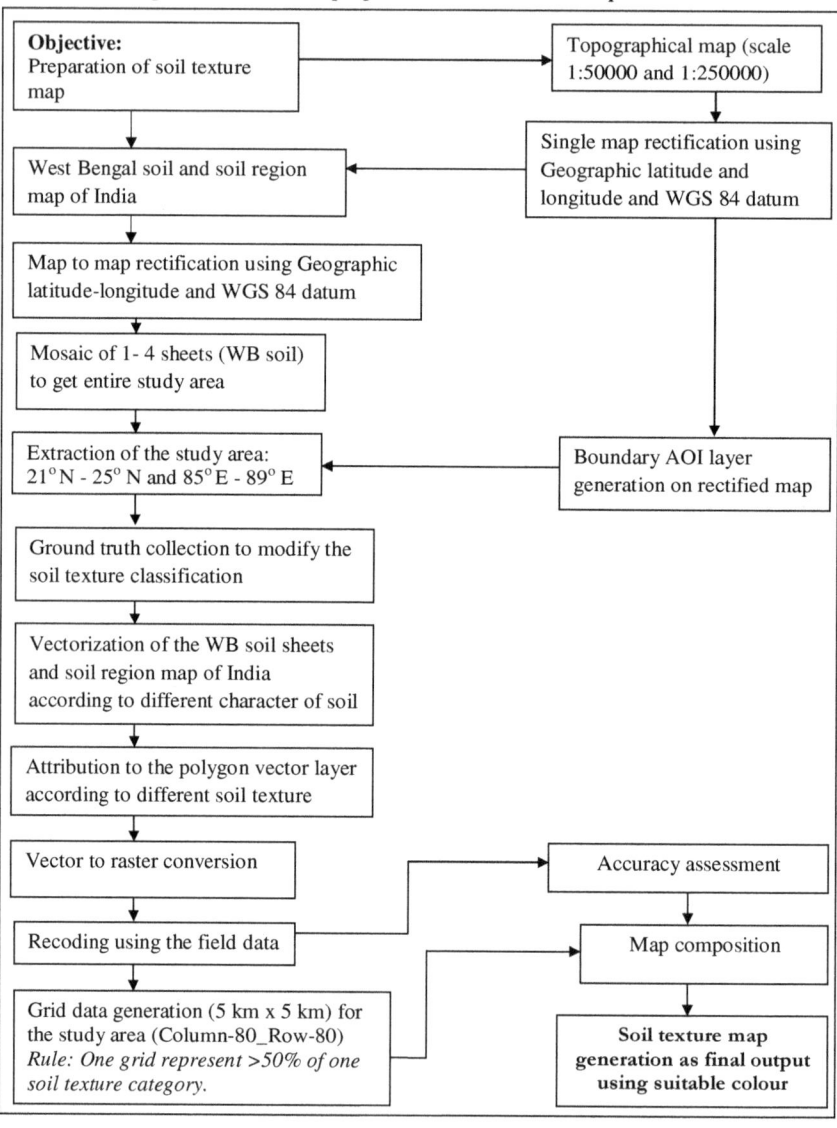

3.2.7 Soil moisture determination

The feel and appearance of the soil indicates soil moisture status. Using a soil tube, soil auger or tile spade, a user can determine moisture content of the soil. User can have an approximate estimation of soil moisture status by firmly squeezing a handful of soil and comparing results with Table 3.11.

Table 3.11. Soil moisture interpretation chart

Soil moisture deficiency	Moderately coarse texture	Medium texture	Fine and very fine texture
0% (field capacity)	Upon squeezing, no free water appears on soil but wet outline of ball is left on hand.		
0-25%	Forms weak ball, breaks easily when bounced in hand.*	Forms ball, very pliable, slicks readily.*	Easily ribbons out between thumb and forefinger.*
25-50%	Will form ball, but falls apart when bounced in hand.*	Forms ball, slicks under pressure.*	Forms ball, will ribbon out between thumb and forefinger.*
50-75%	Appears dry, will not form ball with pressure.*	Crumbly, holds together from pressure.*	Somewhat pliable, will ball under pressure.*
75-100%	Dry, loose, flows through fingers.	Powdery, crumbles easily.	Hard, difficult to break into powder.

*Squeeze a handful of soil firmly to make ball test   *Source Reference: Miles D L 1999*

The available moisture content depends greatly on the soil texture and structure. A range of values for different types of soil is given in the following Table 3.12.

Table 3.12. Usable soil moisture capabilities

Texture	Available moisture cm/m
Fine and very fine (clay, silty clay, sandy clay, silty clay loam, clay loam)	13.32 - 20.83
Medium (silt loam, sandy clay loam, loam, very fine sandy loam)	11.68 - 20.01

| Moderately coarse (fine sandy loam, sandy loam) | 8.33 – 13.32 |

The field capacity, permanent wilting point and available water content are called the soil moisture characteristics. They are constant for a given soil, but vary widely from one type of soil to another.

3.3 Result and discussion

LANDSAT 7, ETM+ satellite imagery are used to determine the land use/land cover of the study area with the help of pixel by pixel standard classification methodology. Three type of data set are generated with the grid size of 30 m, 1 km and 5 km. Overall 90% accuracy is calculated following the accuracy assessment procedure for 30m gridded land use/land cover data set. The western part of the study area is relatively more heterogeneous having forested area, crop land and shrub land etc. Therefore the climatic variables patterns are expected to be more complex.

Surface cover is very important for governing air temperature differences in the area (Eliasson 1992). Surface properties, such as roughness and vegetation play a dominant role in determining surface atmosphere energy exchanges. In most cases, observed air temperatures varied significantly according to ambient land cover types, and air temperatures decreased as the amount of vegetated area around the sites increased. These land cover effects are slightly stronger at night than during the day, and they weakened as the amount of cloud cover or the wind speed increased, especially at night (Yokobori and Ohta 2009). The urban area increases the absorption and storage of incoming solar radiation because the canyon geometry of an urban area increases the surface area that leads to multiple reflections thus increasing effective absorption. The characteristics of urban surfaces can result in a reduction of evapo-transpiration. In addition, urban areas are sources of anthropogenic heat (Oke 1982). As a result, the energy balance in an urban area is disturbed. This effect of the urban climate on air temperatures is called urban heat island (UHI), which refers to the temperature difference between urban areas and surrounding cool rural areas (Yokobori and Ohta 2009). Vegetation can play an important role in moderating this phenomenon, because the incoming radiation energy in a vegetated area is converted into latent heat during the day and the sensible heat flux is lower than that of a built-up area covered by asphalt or concrete (Jonsson 2004). The shading of vegetation reduces the radiant absorption of the ground. At night, a vegetated area with short plants or an area of open bare ground may cool rapidly due to its higher radiative cooling rate (Spronken-Smith and Oke 1999). Many factors are responsible for air temperature differences. If one focuses on land cover, this is essential to understand that- air temperature differences in the study area, because land cover directly affects the energy balance characteristics.

NBSS & LUP soil sheets 1 to 4 are used to generate the soil texture database of the study area with the help standard GIS methodology. All the soils are categorized into nine (9) soil texture classes according to their characteristics, namely loamy sand, sandy loam, silty loam, sandy clay loam, clay loam, silty clay loam, clay loam, sandy clay, and silty clay. The eastern and southern parts are relatively more heterogeneous then western and northern parts. Therefore the climatic variables patterns are expected to be more complex.

The air temperature also influenced by the soil texture, soil moisture and soil temperature (Yeh et al. 1984). Soil moisture is an important factor influencing soil temperature. Average monthly soil temperature near the soil surface is calculated using equations developed by Parton (1984). The equations calculate maximum soil temperature as a function of the maximum air temperature and the canopy biomass (lower for high biomass) while the minimum soil temperature is a function of the minimum air temperature and canopy biomass (higher for higher biomass). The actual soil temperature used for decomposition and plant growth rate functions is the average of the minimum and maximum soil temperatures. Changes in soil temperature will lag behind the changes in air temperature if the difference in air temperature between months is greater than 2° C.

CHAPTER 4

TOPOGRAPHIC CHARACTERISTICS

CHAPTER OUTLINE

- Approaches to studying topography
- Forms of topographic data
- Digital elevation model
- Data uses for digital elevation model map preparation
- Methodology for preparation of digital elevation model map
- Surface analysis using DEM
- Result and discussion

4.0 TOPOGRAPHIC CHARACTERISTICS

Topography is the study of Earth's surface shape and features or those of other planets, moons, and asteroids. It is also the description of such surface shapes and features, especially their depiction in maps. In a broader sense, topography is concerned with local detail in general, including not only relief but also vegetative and human-made features, and even local history and culture. Topographic maps with elevation contours have made "topography" synonymous with relief. Topography specifically involves the recording of relief or terrain, the three-dimensional quality of the surface, and the identification of specific landforms. This is also known as geo-morphometry. In modern usage, this involves generation of elevation data in electronic form. It is often considered to include the graphic representation of the landform on a map by a variety of techniques, including contour lines, hypsometric tints, and relief shading.

4.1 Approaches to studying topography

There are a variety of approaches to studying topography, which method(s) are used depend on the scale and size of the area under study, its accessibility and the quality of existing surveying processes.

- *Direct survey-* Surveying helps determine accurately the terrestrial or three-dimensional space position of points and the distances and angles between them using leveling instruments such as theodolites, dumpy levels and clinometers.

- *Remote sensing-* Remote sensing has greatly speeded up the process of gathering information, and has allowed greater accuracy control over long distances, the direct survey still provides the basic control points and framework for all topographic work, whether manual or GIS-based.

- *Stereo-Photogrammetry-* It is a measurement technique for which the co-ordinates of the points in 3D of an object are determined by the measurements made in the overlapping portions in two photographic images taken from different positions, usually from different exposure stations of an aerial photography flight. In this technique, the common points are identified on each image. A line of sight can be built from the camera location to the point on the object. It is the intersection of its rays which determines the relative three-dimensional position of the point. Known control points can be used to give these relative positions absolute values. More sophisticated algorithms can exploit other information on the scene known a priori, for example, symmetries in certain cases allowing the rebuilding of three-dimensional co-ordinates starting from one only position of the camera.

- *Aerial and satellite imagery-* Besides their role in photogrammetry, aerial and satellite imagery can be used to identify and delineate terrain features and more general land-cover features. Certainly they have become more and more a part of geo-visualization, whether maps or GIS systems. False-color and non-visible spectra imaging can also help determine the characteristics of the land by delineating vegetation and other land-use information more clearly. Images can be in visible colors and in other spectra (Pal et al. 1992).

- *Radar and sonar-* Satellite radar mapping is one of the major techniques of generating digital elevation models (Pal et al. 1994), normally using a process called interferometry. Similar techniques are applied in bathymetric surveys using sonar to determine the terrain of the ocean floor. In recent years, light detection and ranging (LIDAR), a remote sensing technique using a laser instead of microwaves, has increasingly been employed for complex mapping needs such as charting canopies and monitoring glaciers.

4.2 Forms of topographic data

Terrain is commonly modeled either using vector (triangulated irregular network or TIN) or gridded (raster image) mathematical models. In the most common applications in environmental sciences, land surface is represented and modeled using gridded models. In civil engineering and entertainment businesses, the most representations of land surface employ some variant of TIN models. In practice, surveyors first sample heights in an area, then use these to produce a digital land surface model (also known as a digital elevation model). Digital land surface models should not be confused with digital surface models, which can be surfaces of the canopy, buildings and similar objects.

- *Raw survey data-* Topographic survey information is historically based upon the notes of surveyors. They may derive naming and cultural information from other local sources (for example, boundary delineation may be derived from local cadastral mapping. While of historical interest, these field notes inherently include errors and contradictions that later stages in map production resolve.

- *Remote sensing data-* As with field notes, remote sensing data (aerial and satellite photography, for example) is raw and uninterrupted. It may contain holes (due to cloud cover for example) or inconsistencies (due to the timing of specific image captures). Most modern topographic mapping includes a large component of remotely sensed data in its compilation process.

- *Topographic map-* Topographic maps demonstrate not only the contours, but also any significant streams or other bodies of water, forest cover, built-up areas or individual buildings (depending on scale), and other features and points of interest. Existing topographic survey maps, because of their comprehensive and encyclopedic coverage, form the basis for much derived topographic work. Digital elevation models, for example, have often been created not from new remote sensing data but from existing paper topographic maps. Many government and private publishers use the artwork (especially the contour lines) from existing topographic map sheets as the basis for their own specialized or updated topographic map

- *Topological model-* Geographic information system (GIS) can recognize and analyze the spatial relationships that exist within digitally stored spatial data. These topological relationships allow complex spatial modelling and analysis to be performed. Topological relationships between geometric entities traditionally include adjacency (what adjoins what), containment (what encloses what), and proximity (how close something is to something else).

i. reconstitute a sight in synthesized images of the ground,

ii. determine a trajectory of over flight of the ground,

iii. calculate surfaces or volumes,

iv. trace topographic profiles,

4.3 Digital elevation model

A digital elevation model (DEM, or more correctly a land surface model - LSM) is one of the most useful sources of information for spatial modelling and monitoring, with applications as diverse as: environment and Earth science, e.g. catchment dynamics and the prediction of soil properties; engineering, e.g. highway construction and wind turbine location optimization; military, e.g. land surface visualization, and; entertainment, e.g. landscape simulation in computer games (Hengl and Evans 2007). The extraction of land surface parameters – whether they are based on 'bare earth' models such as DEMs derived from contour lines and spot heights, or 'surface cover' models derived from remote sensing sources that include tree top canopies and buildings for example – is becoming more common and more attractive due to the increasing availability of high quality and high resolution DEM data (Gamache 2004). The first release of shuttle radar topography mission (SRTM) data was provided in 1-degree digital elevation model (DEM) tiles from the USGS ftp server. The data was released continent by continent, as and when the data was processed by NASA and the USGS. For the United States, data was made available at 1-arc second resolution (approximately 30 m at the equator), but for the rest of the world the 1-arc second product is degraded to 3-arc seconds (approximately 90 m at the equator). SRTM elevation data has now been released for the entire terrestrial surface, and a "finished" product has now been released. The data is projected in a Geographic (lat/long) projection, with the WGS84 horizontal datum and the EGM96 vertical datum.

4.4 Data uses for digital elevation model map preparation

Different types of data are used for preparation of elevation map of the study area, like shuttle radar topography mission (SRTM) data and topographical map. One of the most widely used digital elevation model (DEM) data sources is the elevation information provided by the shuttle radar topography mission (SRTM) (Coltelli et al. 1996; Gamache 2004), but as with most other DEM sources, the SRTM data requires significant levels of pre-processing to ensure that there are no spurious artifacts in the data that would cause problems in later analysis such as pits, spikes and patches of no data (Dowding et al. 2004; Gamache 2004; Chaplot et al. 2006; Fisher and Tate 2006). In the case of the SRTM data, these patches of no data (USGS 2006b) are filled, preferably with auxiliary sources of DEM data, like topographical maps are used for our study. All other details of the data are given in the Table 4.1.

Table 4.1. Sources, year of publication and scale of SRTM data and topographical maps

Sl. No.	Name of materials	Scale/ Resolution	Year of publication	Source
1	Shuttle radar topography mission (SRTM) data	3-arc seconds	2003	ftp://e0srp01u.ecs.nasa.gov
2	Topographical map	1:250000	1960	University of Texas Libraries, Austin
3		1:50000	1973 - 1980	Survey of India, Kolkata

4.5 Methodology for preparation of digital elevation model map

4.5.1 SRTM Data processing methodology

The method described by Reuter et al. (2007) is adopted for the SRTM data processing. Firstly it involves importing and merging the 1-degree tiles into continuous elevation surfaces in Arc GRID format. The second process fills small holes iteratively, and the cleaning of the surface to reduce pits and peaks. The third stage then interpolates through the holes using a range of methods. The method used is based on the size of the hole, and the landform that surrounds it. Arc/Info AML model is used to processing the SRTM data. The original SRTM DEM (finished grade data from ftp://e0srp01u.ecs.nasa.gov) is used to produce contours or points. Processing is made on a void by void basis. In cases when a higher resolution auxiliary DEM is available, point coverage is produced of the elevation values at the centre of each cell of the auxiliary DEM within void areas. When no high resolution auxiliary DEM is available, the 30 second SRTM DEM is used as an auxiliary for large voids. For areas with a high resolution auxiliary DEM: the contours and points surrounding the hole and inside the hole are interpolated to produce a hydrologically sound DEM using the TOPOGRID algorithm in Arc/Info. TOPOGRID is based upon the established algorithms of Hutchinson (1988; 1989), designed to use contour data (and stream and point data if available) to produce hydrologically sound DEMs. This process interpolates through the no-data holes, producing a smooth elevation surface where no data was originally found. Drainage enforcement is activated, and the tolerances set at 5 for "tolerance 1", representing the density and accuracy of input topographic data, and a horizontal standard error of 1.0 m and vertical standard error of 0.1 m.

4.5.2 Topographical map data processing methodology

At first we perform single map rectification using the geographical co-ordinate system and WGS 84 datum with a RMS error of 0.02. ERDAS IMAGINE 9.0 is used to rectify topographical maps. The entire study area does not appear in single sheet. All the geo-referenced topographical map sheets of the area mosaicked to get the entire study area. Using the similar projection as well as datum for all the soil maps mosaicked in the ERDAS IMAGINE. The study area ($21°$ N - $25°$ N and $85°$ E - $89°$ E) extracted by sub-setting from the mosaic map using the coordinates of four corners. To generate the digital elevation model of the study area ArcInfo and ERDAS IMAGINE s/w are used. First create all contours vector layer (arc coverage) by digitization process in the ArcInfo and ERDAS IMAGINE s/w environment. With the help of polyline tools different soil texture area are digitized. Then 'CLEAN' and 'BUILD' are performed to remove digitization error (Longley et al. 2005). According to the topographical map contour all the data are introduced into the attribute table of the line/arc coverage layer against each Polyline.

Another point layer is generated in the same s/w environment and the collected spot height according to the longitudinal and latitudinal value. Then spot height attribute for each sample location is introduced to the point layer. Arc coverage, which represents the contours, is used to build 3D surfacing in the ERDAS IMAGINE s/w environment. All point data are interpolated to the raster through interpolation techniques in the ArcGIS s/w environment. The best interpolations methods can be generalized as: kriging or Inverse Distance Weighting interpolation for small and medium size voids in relatively flat low-lying areas; spline interpolation for small and medium sized voids in high altitude and dissected terrain; triangular irregular network or inverse distance weighting interpolation for large voids in very flat areas, and an advanced spline method for large voids in other terrains.

Prepared data sets in this process are compared with the SRTM data set and the accuracy of both data sets is ascertained. The interpolated DEM data is then merged with the original DEM to provide continuous elevation surfaces without no-data regions. This entire process is performed for tiles with large overlap with neighboring tiles, thus ensuring seamless and smooth transitions in topography in large void areas. The resultant seamless dataset is then clipped along coastlines using the shorelines and water mask.

4.5.3 Output generation

After obtaining satisfactory level of accuracy, generation of the final output is accomplished. The entire data range is classified into 11 groups with the interval of 100 m altitude. Two sets of data/map are generated to represent the digital elevation model, as first one with 90 m (Figure 4.1), second one with 1km (Figure 4.2) and third one with 5km spatial resolution (Figure 4.3). A contrasting color look up' table (LUT) is developed in the computer for each class by assigning desired RGB against the code (Table 4.2). A digital elevation map is thereby generated wherein individual categories get distinct color.

Table 4.2. Recode values for different elevated classes

Row No.	classes (Elevation in m)	Recode Value	RGB Color presentation
1	Less Than 0	0	0.043: 0.173: 0.478
2	0 - 100	1	0.078: 0.380: 0.529
3	100 – 200	2	0.125: 0.588: 0.573
4	200 – 300	3	0.098: 0.690: 0.431
5	300 – 400	4	0.031: 0.812: 0.149
6	400 –500	5	0.251: 0.890: 0.000
7	500 – 600	6	0.722: 0.961: 0.000
8	600 – 700	7	0.980: 0.918:0.020
9	700 – 800	8	0.949: 0.741: 0.059
10	800 – 900	9	0.910: 0.584: 0.102
11	900 – 1000	10	0.831: 0.427: 0.173
12	More Than 1000	11	0.761: 0.322: 0.235

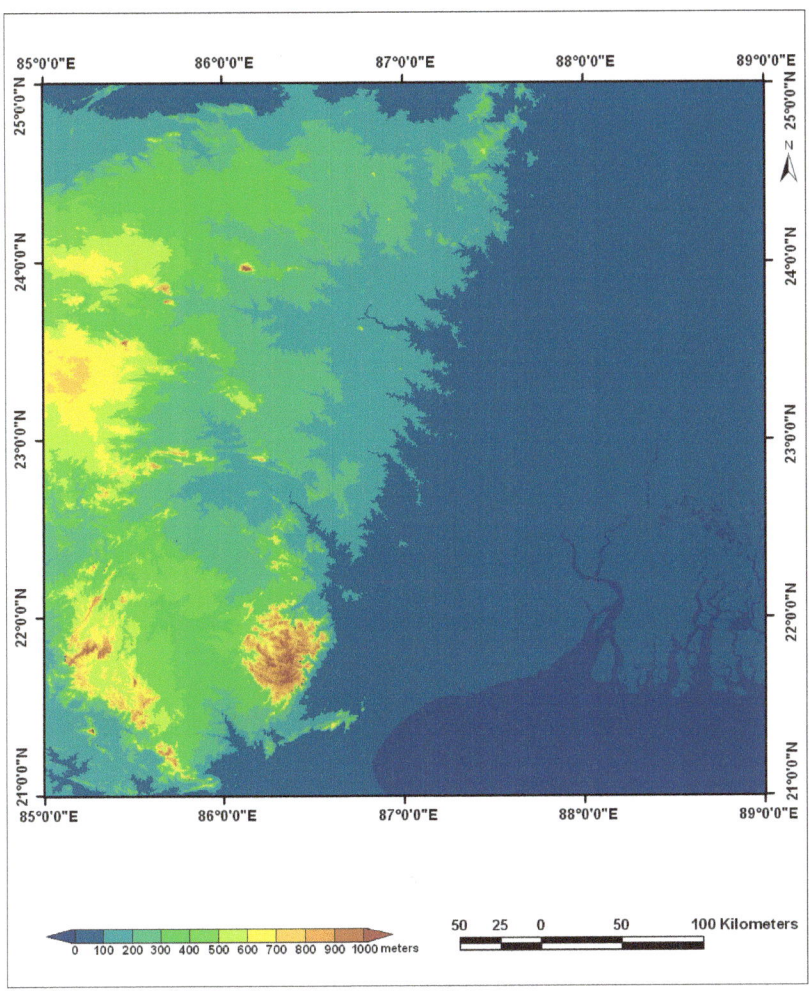

Figure 4.1. Digital elevation model of the study area (90 m x 90 m), generated from SRTM data and topographic maps

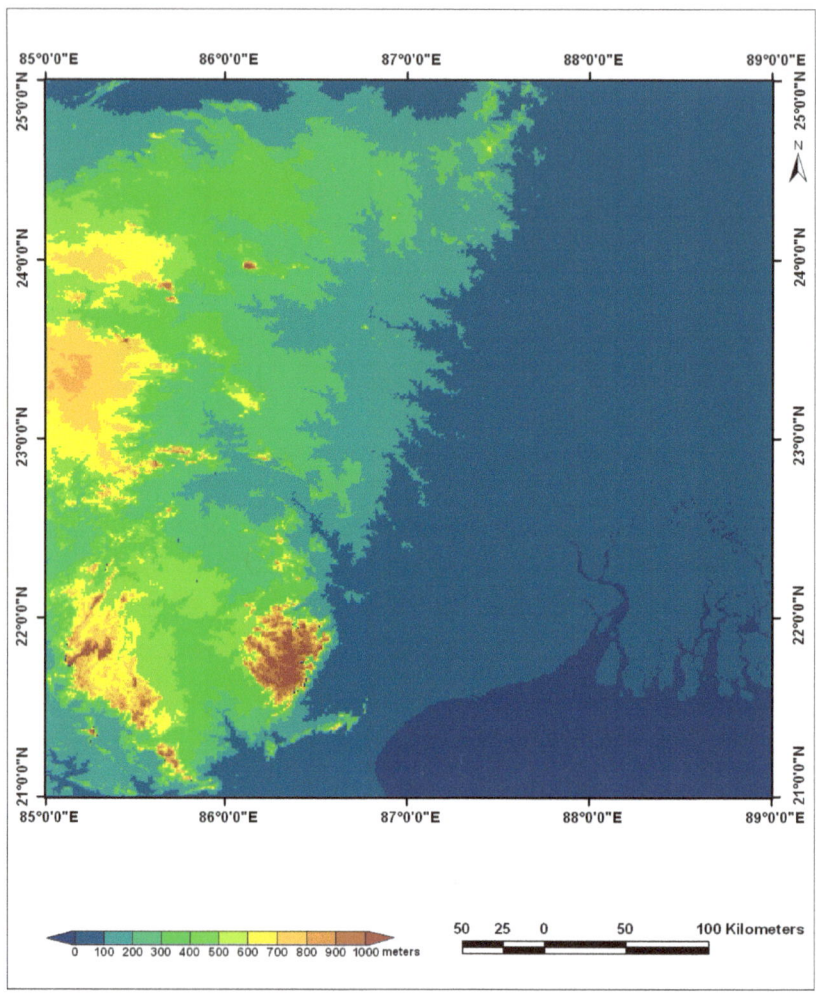

Figure 4.2. Digital elevation model of the study area (1 km x 1 km), generated from SRTM data and topographic maps

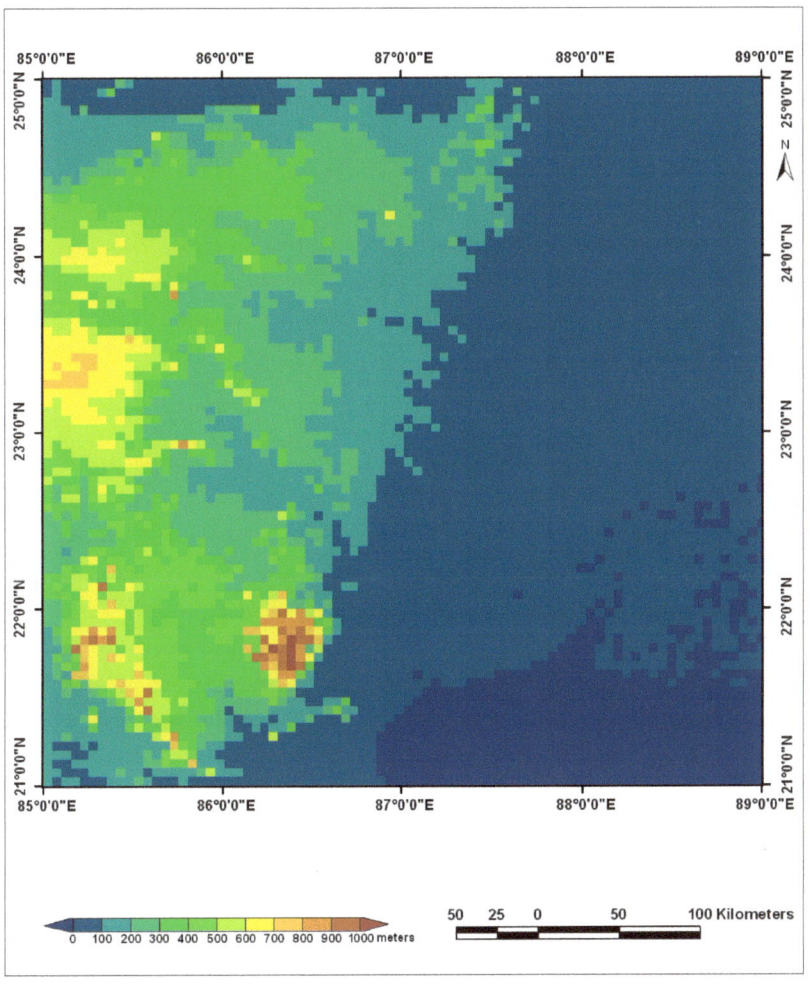

Figure 4.3. Digital elevation model of the study area (5 km x 5 km), generated from SRTM
data and topographic maps

4.7 Result and discussion

SRTM and topographical maps are use to generate the DEM with the help of surfacing techniques in the GIS environment. Using the 3-D analysis process the digital elevation model is generated from contours, which have been digitized from topographical maps. Three sets of data are generated to represent the digital elevation model, as first one with 90 m , second one with 1 km and finally with 5 km spatial resolution as shown in Figure 4.1, 4.2 and 4.3. The elevation of the eastern and southern parts of the study area is less than 100 m, the slope of the topography also is very less, while the south-west parts relatively more elevated (>1000 m) and having high slope than eastern, north-east and southern parts of the study area.

In the troposphere the air temperature regularly decreases with an increase in elevation. The decrease in temperature with elevation is called the environmental lapse rate of temperature *or* normal lapse rate of temperature. On an average the normal lapse rate of temperature is 0.65° C every 100 meter. The environmental lapse rate of temperature is the actual vertical change in temperature on any given day and can be a little greater or less than 0.65° C / 100 meters. Moreover, it may be noted that the decrease in temperature with height is caused by increasing distance from the source of energy that heats the air, i.e. the Earth's surface. Air is warmer near the mean sea level because it's closer to its source of heat plus a highest concentration of greenhouse gases (hear absorbing entities) exists near the mean sea level. The further away from the surface, the cooler the air will be. It's like standing next to a fire; the closer we are the warmer we feel.

The lapse rate is defined as the rate of decrease with height for an atmospheric variable. In general, a lapse rate is the negative of the rate of temperature change with altitude change, thus:

$$\gamma = -\frac{dT}{dz}$$

Where γ is the lapse rate given in units of temperature divided by units of altitude, T = temperature, and z = altitude.

CHAPTER 5

LAND SURFACE TEMPERATURE ASSESSMENT

CHAPTER OUTLINE

- Location of the study area
- Calculating land surface temperature
- Results and discussion

5.0 LAND SURFACE TEMPERATURE ASSESSMENT

The wide use of digital technique for image analysis is mainly because of the amenability of digital data through computer assisted techniques. In the present study, the ERDAS IMAGINE 9.0 system is used to perform a number of digital image enhancement technique & digital image processing. These techniques include principal component analysis, band rationing, image subtraction, various stretching techniques of the DN values, addition (band combination) for generating FCC derivatives, and conversion from digital number, NDVI & temperature modelling. Surface temperature modelling is attempted using satellite imagery (Vogt et al. 1997). LANDSAT-7, ETM+ data is used to identify and assess areas affected by massive coal burning in a thermal power station. Kolaghat thermal power station (KTPS) and its surroundings areas are selected for this work. It is located in the northern part of Purba Medinipur district of West Bengal, India.

5.1 Location of the study area

The study area lies between 87° 45′ E, 21° 20′ N and 88° E, 22° 33′ N. Kolaghat thermal power station (KTPS) is located in between the Kolaghat and Mecheda stations of south eastern railway. It is situated on the right bank of the river Rupnarayan. It falls under CD block Panskura-II in Purba Medinipur districts, West Bengal, India (Figure 5.1). It is bounded in the east by Rupnarayan River, to the north national highway no 6 runs in east west direction and the south eastern railway is extended through the area. It is bounded in the west by the NH 41. The southern mouza i.e. the Shantipur Mouza is located to the south of the KTPP while the Rakshachak mouza is located to the north of the power plant.

5.2 Calculating land surface temperatures

This model is used to calculate the land surface temperature from band 7 (ETM+). ERDAS IMAGINE 9.0 is used for this purpose. Unlike the published TM-temperature model, we considered here the min-max digital count for every image. First, atmospheric correction is applied to the LANDSAT-7, ETM+ data. Next the surface temperatures of the area are computed grid-wise. In relation of temperature assessment the spectral radiance temperature of each pixel is derived first. Then the brightness temperature is calculated from spectral radiance temperature of each pixel. NDVI, fractional NDVI, and emissivity are calculated step by step (Samanta 2009). Finally the land surface temperatures are calculated using the brightness temperature and emissivity results. These land surface temperatures help to find out the adverse impact of coal fire zone in the study area.

5.2.1 Conversion from digital number (image) to radiance

ETM+ bands are quantized as 8 bit data; all information is stored in digital number (DN) with range between 0 and 255 (8 bit). The data was converted to radiance using a linear equation (Sobrino 2005) as shown below-

$$CV_R = G(CV_{DN}) + B \text{ -------------- (Equation 1)}$$

Where:

CV$_R$ is the cell value as radiance
CV$_{DN}$ is the cell value digital number
G is the gain (0.05518 for ETM+ 7)
B is the offset (1.2378 for ETM+ 7)

5.2.2 Conversion from radiance to brightness temperature

The thermal band's radiance values are converted to brightness temperature values by applying the inverse of the Planck function. The formula used (Sobrino et al. 2004) to convert radiance to brightness temperature is as following-

$$T = \frac{K_2}{\ln(\frac{K_1}{CV_R} + 1)} \text{ -------------- (Equation 2)}$$

Where:
  T is degrees Kelvin
  CV_R is the cell value as radiance
  K_1 is calibration constant 1{666.09 (for ETM+)}
  K_2 is calibration constant 2{1282.71 (for ETM+)}

5.2.3 Conversion from digital number (image) to NDVI

NDVI for each pixel is calculated using the following equation-

$$NDVI = \frac{r_{NIR} - r_R}{r_{NIR} + r_R} \text{ -------------- (Equation 3)}$$

Where:
  r_{NIR} and r_R represent the percentage reflected radiation in the near-infrared and red portion of the spectrum respectively.

5.2.4 Conversion from NDVI to emissivity

In general, the effective emissivity of a pixel should be estimated by summing up the contributions from its surface type (Sobrino et al. 2001). Van de Griend et al. (1993) found a high correlation between measured **e** and NDVI. The following equation describes the relation between measured e and NDVI-

$$e = a + b \text{ x } \ln (NDVI) \text{ -------------- (Equation 4)}$$

Where:
  a is 1.0094
  b is 0.047

5.2.5 Land surface temperature (LST) retrieval

LST is derived from ETM+7 using model developed by Sobrino et al. (2004) and Jackson et al. (2004) which use spectral surface emissivity and brightness temperature values of the particular scenes.

$$St = \frac{T}{1+(\lambda x T / \rho)\ln \varepsilon} \text{--------------- (Equation 5)}$$

Where,
S_t = Land Surface Temperature
λ is wavelength of emitted radiance (λ = 11.5 µm),
ρ = h×c/σ (1.438 ×10^{-2} m K),
T = Brightness temperature values
ε = Emissivity

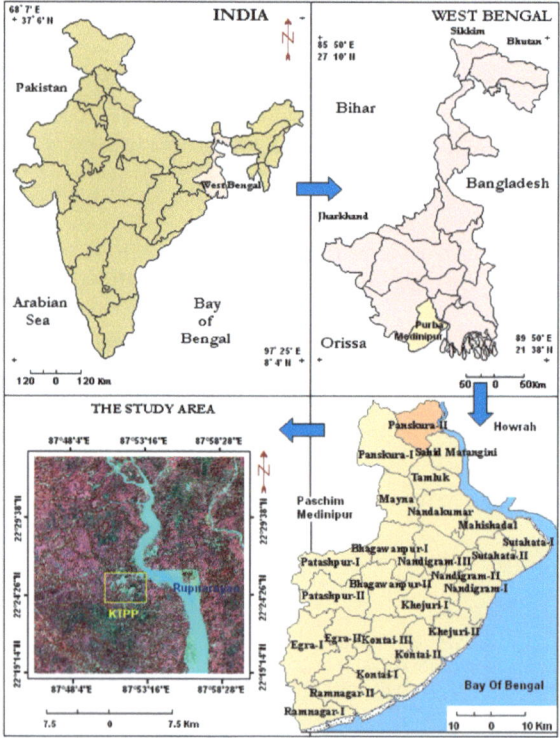

Figure 5.1. Location map of the Kolaghat Thermal power plant Region

5.3 Results and discussion

In the model maker of ERDAS IMAGINE 9.0 these 5 equations are processed to calculate the land surface temperature. In the model output we have derived a raster matrix with temperature as a cell value. The calculated land surface temperature shows a range of temperature varying from 26 to 40 degree centigrade. The hot spots are clearly identified from the resultant image. The KTTP region (the thermal plant) and the brick kiln industry regions show the high temperature value for their thermal influencing properties with an ostensibly high emittance TIR radiation.

Through the thermal modelling (using the thermal bands) the hot spots are easily indentified. This indicates the LANDSAT ETM+ thermal infrared data can be utilised to calibrate reasonably accurate brightness temperatures if appropriate ground truth and ancillary data are available.

CHAPTER 6

SPATIAL INTERPOLATION AND CLIMATOLOGICAL MODELLING

CHAPTER OUTLINE
- Spatial interpolation of climate data
- Climatological modelling of temperature and rainfall
- GIS methodology for the modelling
- Results and verification analysis

6.0 SPATIAL INTERPOLATION AND CLIMATOLOGICAL MODELLING

6.1 High resolution data set of surface climate

Spatially inclusive representations of surface climate are required for many purposes in applied and theoretical environmental science. Examples include biogeochemical modelling (Cramer and Fischer 1996), forestry (Booth and Jones 1998), agriculture (Changnon and Kunkel 1999), hydrology and water resources (Arnell 1999), climate change studies (Hulme and Jenkins 1998; Hulme et al. 1999; Giorgi and Francisco 2000; Das and Lohar 2005). Typically, the required spatial resolution of climate data increases with the resolution of analysis.

6.1.1 Data source

Global data sets of monthly time series of precipitation (Eischeid et al. 1991), mean temperature (Jones 1994), maximum and minimum temperature (the global historical climatology network; Easterling et al. 1997) for several thousand stations worldwide are searched for additional stations. Climate Research Unit (CRU) also holds smaller data sets of monthly time series of the other variables that are used for this study. These data have been quality-controlled and checked for homogeneity. 'Station means' for 1901 to 2002 are calculated from these time series data set. Despite these data collation efforts, the CRU data in many regions still represent only a sub-set of the potentially available stations. NASA Surface meteorology and Solar Energy (SSE) also provides 0.5 degree climate data sets from 1972 to 2002, which are used for this research work.

6.1.2 Data set

The climate stations data are used to construct the climatology represent the fruits of a data collation program at the CRU. The climatology comprises a suite of variables such as mean temperature, diurnal temperature range, relative humidity, sunshine, ground-frost frequency etc. and among all the variables, precipitation and temperature are the most widely available entities, followed by diurnal temperature range (simply the difference between mean maximum and minimum temperature). Precipitation is the densely reported variable with 27000 stations while the coefficient of variation (CV) of precipitation is available for 22000 stations. Differences in temperature measurement timings have been shown to induce disparities of several tenths of a degree Celsius (Karl et al. 1986; Andersson and Mattison 1991), and different countries calculate mean temperature in various ways, where possible, mean temperature is defined as the average of mean maximum and minimum temperature, which are measured more uniformly across the world. At stations where only mean temperature is available, these values are used, despite the uncertainty about their derivation. Wet-day frequencies are generally expressed as the number of days per month with precipitation > 0.1 mm per month. The data set contains cloud cover and sunshine: Sunshine recordings are supplied as either mean hours per month or percent of maximum possible bright sunshine. Humidity in the CRU data set comprised roughly equal numbers of relative humidity (RH) and vapour pressure *(e)*. Different variables with their nomenclature and unit of CRU data are shown in a tabular form (Table 6.1) as follows.

Table 6.1. Nomenclature and Units of climate variables of CRU data set

Label	Variable	Units
cld	cloud cover	percentage
dtr	diurnal temperature	degrees Celsius
frs	frost day frequency	days
pre	precipitation	millimetres

rhm	relative humidity	percentage
ssh	sunshine duration	hours
tmp	mean temperature	degrees Celsius
vap	vapour pressure	hecta-Pascals
wet	wet day frequency	days
wnd	wind speed	metres per second

Source: http://www.cru.uea.ac.uk

6.2 Interpretation and analysis of climate data

The meteorological data is on a 1 degree longitude by 1 degree latitude equal-angle grid covering the entire globe. Bilinear interpolation is used to produce 10 minutes data set by CRU and 30 minutes data set by NASA SSE. In general, meteorology and solar radiation for SSE Release 6.0 are obtained from the NASA science mission directorate's satellite and re-analysis research programs. Parameters based upon the solar and/or meteorology data are derived and validated based on recommendations from partners in the energy industry. Release 6.0 extends the temporal coverage of the solar and meteorological data from 10 years to more than 22 years (e.g. July 1983 through June 2005) with improved NASA data, and includes new parameters and validation studies.

6.2.1 Accuracy of the climate data

It is generally considered that the quality of measured data is more accurate than satellite-derived data set. However, measurement uncertainties from calibration drift, operational uncertainties, or data gaps are unknown for ground site data sets. In 1989, the world climate research program estimated that most routinely available ground observations had "end-to-end" uncertainties from 6 to 12%. Specialized high quality research sites are hopefully more accurate by a factor of two. SSE estimates are compared with ground site data on a global basis. Meteorological parameters are compared with data from the national climate data center (NCDC) (Table 6.2).

Table 6.2. Linear least squares regression analysis of SSE (Surface meteorology and solar energy) versus NCDC (National Climate Data Center) monthly averaged values (1983 to 2006)

Parameter	Slope	Intercept	R^2	RMSE	Bias
Temperature maximum (° C)	0.99	1.58	0.95	3.12	1.83
Temperature minimum (° C)	1.02	0.10	0.95	2.46	0.24
Temperature average (° C)	1.02	0.78	0.96	2.13	0.58
Temperature diurnal (° C)	0.96	0.80	0.95	2.46	1.07
Relative humidity (%)	0.79	12.72	0.56	9.40	1.92

Source: NASA Surface meteorology and Solar Energy (*http://eosweb.larc.nasa.gov*)

6.2.2 Monthly average meteorological parameters

The published global CRU TS 2.1 dataset with the spatial resolution of 30 minute is used for interpretation of the climate characteristics of the study area. More than fifty samples are collected to cover the total study area. The general interpretation is achieved using some of the following data sets in some selected sample points (Table 6.3 and Figure 6.1).

Average mean temperature and mean diurnal temperature data are collected from CRU data set. Using these two temperatures the average maximum and average minimum temperature are calculated using the simple equations as follows:

$$t_{max} = t_{mean} + 0.5 * dtr$$

$$t_{min} = t_{mean} - 0.5 * dtr$$

Where, dtr =Mean diurnal temperature, t_{max} = average maximum temperature, t_{min} = Average minimum temperature and t_{mean} = average mean temperature

The general average characteristics of the temperature and rainfall have been shown in the Table 6.3 and also displayed in Figure 6.1. Temperature is appreciably higher in the month of April through May (premonsoon season) and in June-July (first half of monsoon season), demarked as the hottest months in the study area (Figure 6.1) while the bulk of (90% of the year) rainfall occur in the month of June-July-August–September, called rainy season or monsoon season (Figure 6.1). Due to vertical sunshine on the 23.5 degree latitude in the month of June (21 June, the summer solstice in northern hemisphere) the temperature goes up to maximum level and the rainfall starts peaking in the area following that time due to the concomitant advent of monsoon.

Table 6.3. Monthly average meteorological parameters at sample locations [CRU, 1983-2002]

Location Lon/Lat	Month	Derived average max. temperature (°C)	Derived average min. temperature (°C)	Average mean temperature (°C)	Total precipitation (mm)
86° E and 24° N	January	17.65	16.55	17.10	13.6
	February	20.60	19.00	19.80	15.4
	March	25.85	24.35	25.10	16.3
	April	30.65	28.95	29.80	18.0
	May	33.05	29.75	31.40	43.1
	June	35.15	25.65	30.40	193.4
	July	35.35	20.25	27.80	307.9
	August	34.45	20.55	27.50	299.6
	September	32.05	22.35	27.20	243.2
	October	27.10	23.70	25.40	83.2
	November	21.35	20.85	21.10	4.9
	December	17.55	17.05	17.30	3.2
87° E and 23° N	January	19.55	18.65	19.10	11.1
	February	22.75	21.05	21.90	24.1
	March	28.05	25.95	27.00	31.0
	April	32.55	29.45	31.00	43.1
	May	34.55	29.65	32.10	78.0
	June	36.15	25.65	30.90	216.4
	July	36.00	21.40	28.70	285.2
	August	35.70	21.30	28.50	313.8
	September	33.95	22.85	28.40	244.8
	October	29.00	24.60	26.80	95.6
	November	23.10	22.30	22.70	8.9
	December	19.20	18.80	19.00	4.1
86° E and 22° N	January	19.25	18.15	18.70	12.6
	February	22.40	20.40	21.40	32.7
	March	27.10	25.10	26.10	30.8
	April	31.55	28.45	30.00	40.1
	May	34.00	28.60	31.30	79.7
	June	34.90	24.90	29.90	207.7
	July	34.55	20.05	27.30	286.2
	August	34.20	19.60	26.90	337.9
	September	32.55	21.25	26.90	248.2
	October	27.75	23.05	25.40	95.1
	November	22.20	21.40	21.80	12.9
	December	18.90	18.50	18.70	4.4

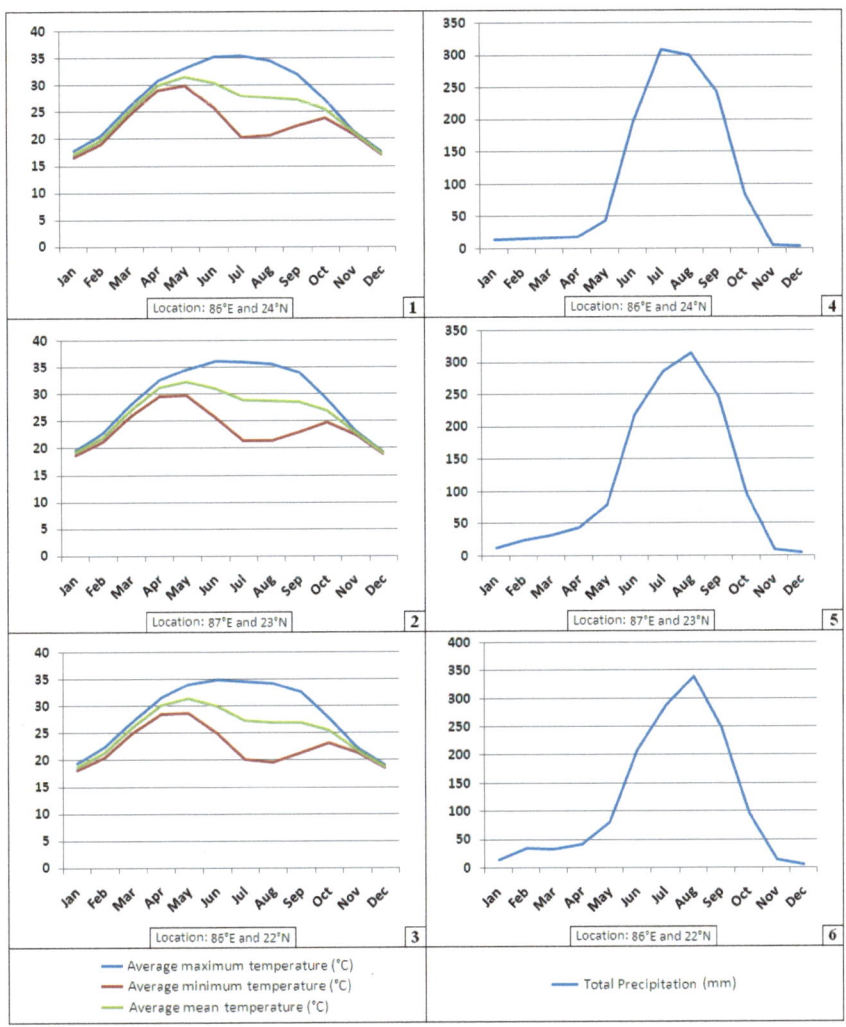

Figure 6.1. Monthly average meteorological parameters of the study area: Left panel (1, 2, 3) displays average maximum, minimum and mean temperature while Right panel (4, 5, 6) for total precipitation of three selected locations.

Table 6.4. Monthly average meteorological parameters at locations-1[NASA SSE, 1983-2002]

Location Lon/Lat	Month	Average max temperature (° C)	Average min temperature (° C)	Average mean temperature (° C)	Total Precipitation (mm)
86° E and 24° N	January	24.67	10.93	17.79	18.38
	February	27.18	13.50	20.33	22.14
	March	32.74	18.24	25.47	19.06
	April	37.14	22.59	29.85	21.35
	May	38.62	25.07	31.84	55.31
	June	35.60	25.41	30.48	188.35
	July	31.02	24.36	27.68	302.33
	August	30.28	24.21	27.23	321.85
	September	30.86	23.72	27.26	220.75
	October	30.51	20.73	25.60	87.25
	November	27.80	15.21	21.49	9.73
	December	25.13	11.48	18.29	3.70

A comparative analysis of the climate variable between CRU data set and NASA SSE data set (Table 6.4) reveals a different character for the temperature and precipitation. The spatial resolution of both dataset is 30 minutes. Same time series is considered for both cases (1983 to 2002). Coming to the first sample point at 86° E and 24° N we can find out the difference of climate variables.

6.3 Interpolation of climate data

Weather and climate remain the dominant components to determine the choice of a particular agriculture activity. Developing countries need an inexpensive and reliable method of interpolation to make good for the lack of adequate number sampling stations for collecting data on the climatic variables that are necessary input for sound agricultural planning. Agricultural research seeks to evolve methods to deal with crop and soil variability, weather generators (computer applications that produce simulated weather data using climate profiles), and spatial interpolation, the estimation of the value of properties at unsampled sites within an area taking cue from the limited number of sampled points (Bouman et al. 1996). Especially in developing countries there is a need for accurate and inexpensive quantitative approaches to spatial data acquisition and interpolation (Mallawaarachchi et al. 1996). Interpolation predicts values for the entire cells in a raster from a limited number of sample data points (Collins et al. 1996). Most data for environmental variables (soil properties, weather) are collected from point sources. It can be used to predict unknown values for any geographic point data. Unknown values are predicted with a mathematical formula that uses the values of nearby known points. The assumption that makes interpolation a viable option, this is spatially distributed objects as well as spatially correlated; in other words, things that are close together tend to have similar characteristics and the similarity keeps fading with distance. For instance, if it is raining on one side of the street, we can predict with a high level of confidence that it is also raining on the other side of the street. On the other hand it would be less certain if it is raining across town and less confident still about the state of the weather in the next county at a considerable distance away. It is easy to see that the values of points close to sampled points are more likely to be similar than those that are farther apart. This is the basis of interpolation. A typical use for point interpolation is to create an elevation surface from a set of sample measurements. By interpolating, the values between the input points will be predicted in a rational way. With the increasing number of applications for environmental data, there is also a growing concern about accuracy and precision. Results of spatial interpolation contain a certain degree of error, and this error is sometimes measurable. Understanding the accuracy of spatial interpolation techniques is a first step toward identifying sources of error and qualifying results based on sound statistical judgments.

6.3.1 Data uses for spatial interpolation

Different types of data are used (Table 6.5) for spatial interpolation process to distribute climate variables in the study area, like CRU data set and topographical map of the corresponding area. (i) CRU of East Anglia University, UK has developed a database of monthly gridded (0.5° x 0.5°) climate observations globally using various techniques (Mitchell and Jones 2005) This data set contains nine climate variables, namely cloud cover, diurnal temperature, frost day frequency, precipitation, relative humidity, sunshine duration, mean temperature, vapour pressure, wet day frequency and wind speed. Diurnal temperature, mean temperature, relative humidity, cloud cover and precipitation variables have been taken for our study. (ii) India Meteorological Department (IMD) has published monthly mean maximum & minimum temperature and monthly total rainfall of important stations (111) in India for the period 1901 to 2000. They have 677 automatic weather stations (AWS) and 1350 automatic rain gauge stations in India. Daily weather phenomena for 24 hour basis are also available from the website (http://www.imd.gov.in/). The IMD data sets are the real ground observation data set. Temperature and rainfall variables have been taken to compare with interpolated CRU data set. (iii) University of Texas Libraries, Austin has published

1:250000 topographical maps in the web which are collected as a collateral data for the geo-referencing purpose. 1:50000 scale topographical maps are also used for the geo-referencing purpose and to generate grid maps to interpolate the climate variables.

Table 6.5. Climate variables and topographical maps for spatial interpolation

Sl. No.	Data sets	Scale/data resolution	Year range	Source
1	Diurnal temperature, mean temperature, relative humidity, cloud cover and precipitation)	30 minute	1901-2002	*http://www.cru.uea.ac.uk*
2				
3	IMD (Temperature and rainfall)	Station/ground data	1901-2000	*http://www.imd.gov.in/*
4	Topographical map	1:250000	1960	University of Texas Libraries, Austin
5		1:50000	1973-1980	Survey of India, Kolkata

It may mentioned herewith that the validity of the CRU data set has been checked through a comparison with the observed available IMD data set as shown in Appendix 7 and 8.

6.3.2 Spatial interpolation

Interpolation predicts values for cells in a raster from a limited number of sample data points. It can be used to predict unknown values for any geographic point data: elevation, rainfall, chemical concentrations, noise levels, and so on. Unknown values are predicted with a mathematical formula that uses the values of nearby known points. Interpolation methods can also be described as "global" or "local." Global techniques (e.g. inverse distance weighted averaging, IDWA) fit a model through the prediction variable over all points in the study area. Typically, global techniques do not accommodate local features well and are most often used for modelling long-range variations. Local techniques, such as splining, estimate values for an un-sampled point from a specific number of neighboring points. Consequently, local anomalies can be accommodated without affecting the value of interpolation at other points on the surface (Burrough 1986). Splining, for example, can be described as deterministic with a local stochastic component (Burrough and McDonnell 1998). A number of usual interpolation techniques are explained as follows:

6.3.2.1 Inverse distance weighted averaging

Inverse distance weighted (IDW) averaging is a method of interpolation that estimates cell values by averaging the values of sample data points in the neighborhood of each processing cell. IDW interpolation explicitly implements the assumption that things that are close to one another are more alike than those that are farther apart. To predict a value for any unmeasured location, IDW will use the measured values surrounding the prediction location. Those measured values closest to the prediction location will have more influence on the predicted value than those farther away. Thus, IDW assumes that each measured point has a local influence that diminishes with distance. It weights the points closer to the prediction location greater than those farther away, hence the name inverse distance weighted (Watson 1985). The surface calculated using IDW depends on the selection of a power value and the neighborhood search strategy. IDW is an exact interpolator, where the maximum and minimum values in the interpolated surface can only occur at sample points. The output surface is sensitive to clustering and the presence of outliers. IDW assumes that the surface is being driven by the local variation, which can be captured through the neighborhood (Hartkamp 1999).

6.3.2.2 Splining

Spline is an interpolation method that estimates values using a mathematical function that minimizes overall surface curvature, resulting in a smooth surface that passes exactly through the input points. Conceptually, the sample points are extruded to the height of their magnitude; spline bends a sheet of rubber that passes through the input points while minimizing the total curvature of the surface. It fits a mathematical function to a specified number of nearest input points while passing through the sample points. This is a deterministic, locally stochastic interpolation technique that represents two dimensional curves on three dimensional surfaces (Eckstein 1989; Hutchinson and Gessler 1994). Splining may be thought of as the mathematical equivalent of fitting a long flexible ruler to a series of data points. There are two types of spline method, they are 'regularized' method and 'tension' method. The regularized method creates a smooth, gradually changing surface with values that can lie outside the sample data range and the tension method controls the stiffness of the surface according to the character of the modeled phenomenon. It creates a less smooth surface with values more closely constrained by the sample data range. Further control of the output surface is accomplished through two additional parameters: (i) weight and (ii) number of points. For the regularized spline method, the weight parameter defines the weight of the

94

third derivative of the surface in the curvature minimization expression. Use of higher weight gives the smoother output surface. The values entered for this parameter must be equal to or greater than zero. The typical values that can be used are 0, 0.001, 0.01, 0.1 and 0.5. For the tension spline method, the weight parameter defines the weight of tension. Higher the weight gives the coarser output surface. The values entered have to be equal to or greater than zero. The typical values are 0, 1, 5, and 10.

6.3.2.3 Kriging

The inverse distance weighted (IDW) and spline methods are referred to as deterministic interpolation methods because they are directly based on the surrounding measured values or on specified mathematical formulas that determine the smoothness of the resulting surface. A second family of interpolation methods consists of geostatistical methods, such as kriging, which are based on statistical models that include autocorrelation, that is, the statistical relationships among the measured points. Because of this, not only do geostatistical techniques have the capability of producing a prediction surface, but they can also provide some measure of the certainty or accuracy of the predictions. The general formula for kriging was developed by Matheron (1970). The most commonly applied form of kriging uses a "semivariogram", a measure of spatial correlation between pairs of points describing the variance over a distance. Weights change according to the spatial arrangement of the samples.

Table 6.6. Comparison of interpolation techniques

Method	Local/ global	Transitions	Interp olator	Limitation	Best For	Output Data structure	Assumption of interpolation model
Classific ation	Global	Abrupt if used alone	No	Delineation of areas and classes may be subjective	Quick assessment	Classifie d Polygon	Homogeneit y within boundary
Trend surface	Global	Gradual	No	Physical meaning of trend may be unclear	Quick assessment and removal of spatial trends	Continu ous gridded surface	Phenomenol ogical explanation of trend

Regression Model	Global with local refinement	Gradual if inputs are Gradual	No	Result depend on the fit of the regression model and the quality and detail of input	Simple numerical modelling of expensive	Polygon or Continuous gridded surface	Phenomenological explanation of regression model
Thiessen Polygon	Local	Gradual	Yes	No error assessment	Nominal data from point observation	Polygon or gridded surface	Best local predictor is nearest data point
Linear Interpolation	Local	Gradual	Yes	No error assessment	Interpolating from point data when densities are high	Gridded surface	Data density are so large that linear approximation is no problem
IDWA	Local	Gradual	No	No error assessment	Quick interpolation from sparse data	Gridded surface	Under laying surface is smooth
Thin Plate Spline	Local	Gradual	Yes	Goodness of fit possible	Quick interpolation of DEM data	Gridded surface, contour lines	Under laying surface is smooth
Kriging	Local with global trends	Gradual	Yes	Error assessment depends on distribution of data & size of interpolate block	Good interpolation from sparse data	Gridded surface	Interpolate surface is smooth

Source: Based on Burrough and McDonnell (1998)

6.4 Methodology of spatial interpolation

The Arc GIS and ERDAS IMAGINE software are used for spatial interpolation of the high resolution climate variables. CRU data sets are selected as a high resolution climate variable. Raster module of Arc GIS and ERDAS IMAGINE are used to rectify the topographical map using the ground control point in the process of single image/map rectification. Geographical coordinate system and WGS84 datum and spheroid is used for this purpose. The vector modules of the two software- Arc GIS and ERDAS IMAGINE are used for creating a grid layer and a point layer for the entire study area. All the climate variables like mean temperature, precipitation, relative humidity and cloud cover data sets are introduced into the attribute table of the point layer. The monthly average and annual measurement of different climate variables are chosen for this purpose. CRU data set for 64 points location are used for the interpolation process. The ArcGIS spatial analyst is used for spatial interpolation of the climate variables. A point data file and a covariate grid are used as inputs to the module. The program yields several output files: (i) a large residual file which is used to check for data errors; (ii) a file that contains an error covariance matrix of fitted surface coefficients; and (iii) an interpolated thematic map showing the distribution of the parameter. All the three interpolation methods, inverse distance weighting, thin plate smoothing splines and kriging are used to interpolate the climate variables (e.g., Appendix 5 and 6). Interpolation of temperature is better handled by splining method than by kriging or inverse distance weighting methods (Table 6.6 and 6.7), because it is faster, easier to use and give more accurate result than other techniques, as also noted in other studies (e.g., Hutchinson and Gessler 1994).

Table 6.7. Average monthly mean temperature (premonsoon) of IMD data set vs. interpolated CRU data set

Station-1	Digha			Station-11	Diamondharbour		
Monthly temperature	March	April	May	Monthly temperature	March	April	May
IMD	26.1	28.4	29.2	IMD	27.2	29.5	30.0
Spline	27.6	29.9	30.9	Spline	27.5	29.8	30.9
Krige	27.7	30.0	31.1	Krige	27.5	29.8	30.9
IDW	27.6	30.2	31.3	IDW	27.5	30.0	31.0
Station-2	Balasore			Station-12	Contai		
IMD	27.3	29.8	30.5	IMD	26.9	29.3	29.5
Spline	27.8	30.3	31.4	Spline	27.5	29.8	30.9
Krige	27.8	30.3	31.4	Krige	27.5	29.9	31.0
IDW	27.8	30.3	31.4	IDW	27.6	30.1	31.2
Station-3	Midnapur			Station-13	Dumdum		
IMD	27.8	31.2	31.4	IMD	27.1	30.1	30.4
Spline	27.6	30.9	32.0	Spline	27.7	30.3	31.3
Krige	27.6	30.9	31.9	Krige	27.7	30.3	31.2
IDW	27.5	30.7	31.9	IDW	27.5	30.2	31.1
Station-4	Ranchi			Station-14	Durgapur		
IMD	24.2	28.1	30.1	IMD	23.4	30.8	31.2
Spline	24.5	28.5	31.0	Spline	26.5	30.6	32.2
Krige	24.5	28.6	31.3	Krige	26.5	30.6	32.2
IDW	24.7	28.8	31.3	IDW	26.5	30.6	32.2
Station-5	Bankura			Station-15	Uluberia		
IMD	29.0	29.7	32.8	IMD	26.9	29.8	30.2
Spline	27.1	30.9	32.3	Spline	27.8	30.4	31.4
Krige	27.1	30.9	32.3	Krige	27.8	30.4	31.3
IDW	27.0	30.8	32.3	IDW	27.8	30.5	31.4
Station-6	Jamshedpur			Station-16	Burdwan		
IMD	27.3	30.9	32.7	IMD	27.5	31.2	32.0

Spline	26.2	30.1	32.2	Spline	27.7	31.1	32.1
Krige	26.2	30.1	32.1	Krige	27.6	31.0	32.0
IDW	26.1	30.0	32.0	IDW	27.5	30.9	31.9
Station-7	Chaibasa			Station-17	Santiniketan		
IMD	26.4	30.4	31.6	IMD	26.3	30.3	30.1
Spline	25.9	29.8	31.9	Spline	27.0	30.9	32.0
Krige	25.9	29.8	31.9	Krige	27.0	30.9	32.0
IDW	25.9	29.8	31.9	IDW	26.9	30.8	31.9
Station-8	Krishnanagar			Station-18	Baripada		
IMD	25.7	29.7	30.0	IMD	27.5	31.1	31.1
Spline	27.0	30.6	31.5	Spline	26.6	29.4	30.8
Krige	27.1	30.6	31.5	Krige	26.7	29.6	31.1
IDW	27.0	30.6	31.5	IDW	26.9	29.8	31.2
Station-9	Purilia			Station-19	Haldia		
IMD	25.6	29.7	30.9	IMD	27.0	29.3	29.7
Spline	26.1	30.1	32.5	Spline	27.4	29.6	30.8
Krige	26.1	30.2	32.4	Krige	27.4	29.6	30.9
IDW	26.0	30.1	32.3	IDW	27.4	29.7	30.8
Station-10	Alipur			Station-20	Hazaribag		
IMD	27.6	30.3	30.8	IMD	23.3	27.9	29.9
Spline	27.7	30.2	31.2	Spline	24.0	28.4	31.2
Krige	27.7	30.1	31.1	Krige	24.1	28.5	31.4
IDW	27.6	30.2	31.2	IDW	24.3	28.7	31.4

Note:
IMD: Observed temperature data set of IMD,
Spline: Interpolated CRU data set using Spline method,
Krige: Interpolated CRU data set using Kriging method
and IDW: Interpolated CRU data set using Inverse distance Weighted method

Different raster data sets (for all the months, seasons and annual) are generated by the spline interpolation process using climate variables. A contrasting color look up' table (LUT) is developed in the computer for each class by assigning desired RGB against the code. Different thematic maps are thereby generated (Figure 6.2, 6.3, 6.4 and 6.5) wherein individual categories get distinct color.

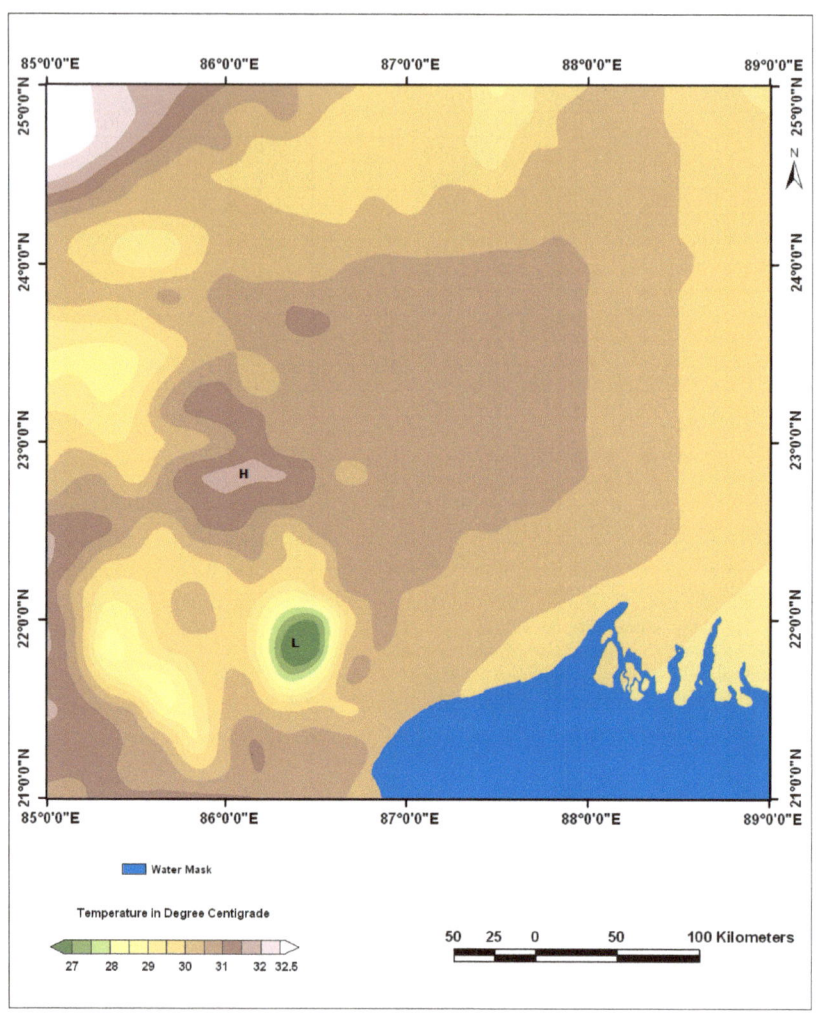

Figure 6.2. Interpolation map of average mean temperature of June (30 years average)

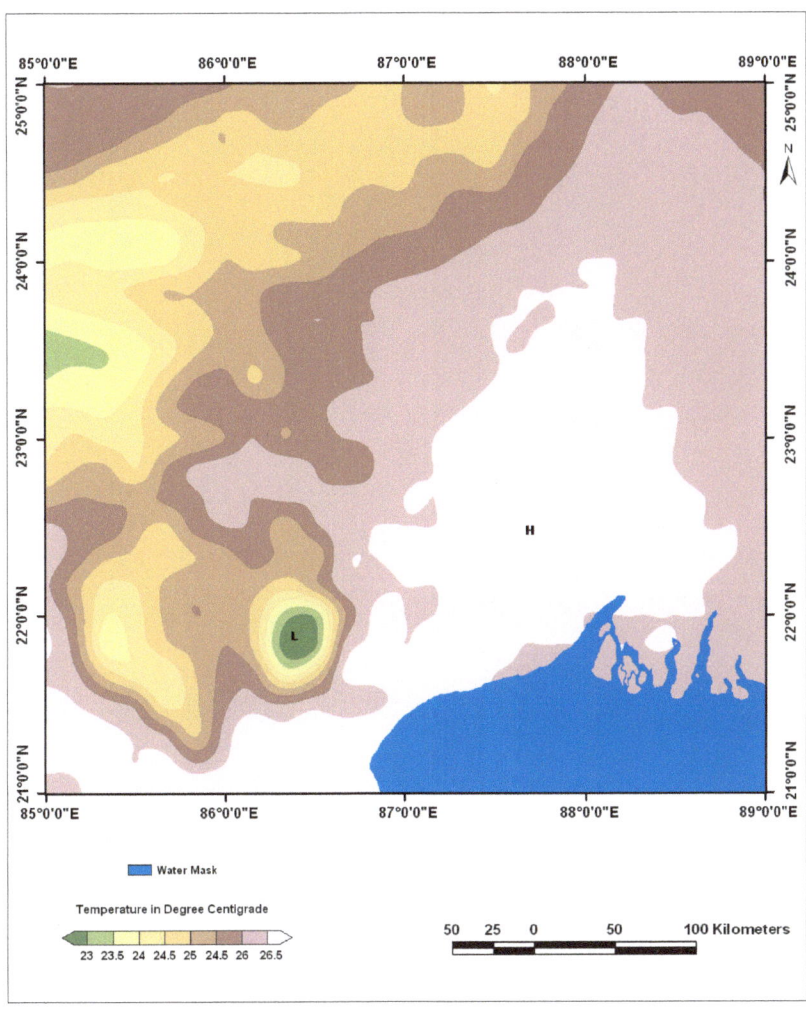

Figure 6.3. Interpolation map of annual average mean temperature (30 years average)

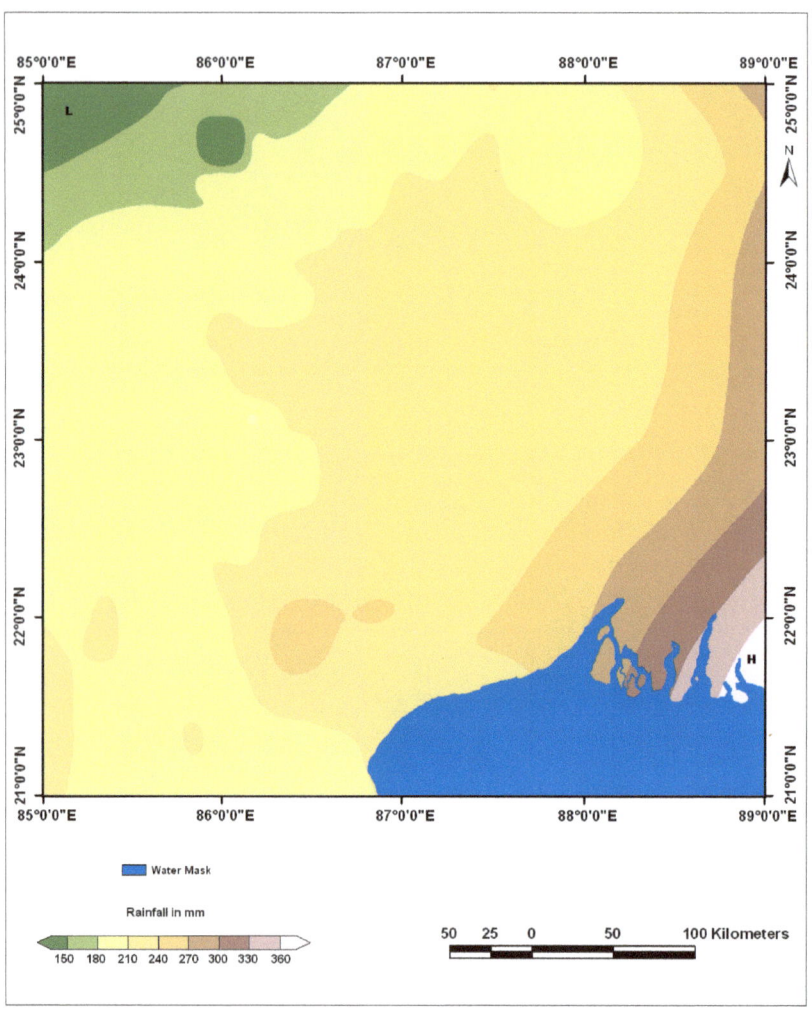

Figure 6.4. Interpolation map of total rainfall of June (30 years average)

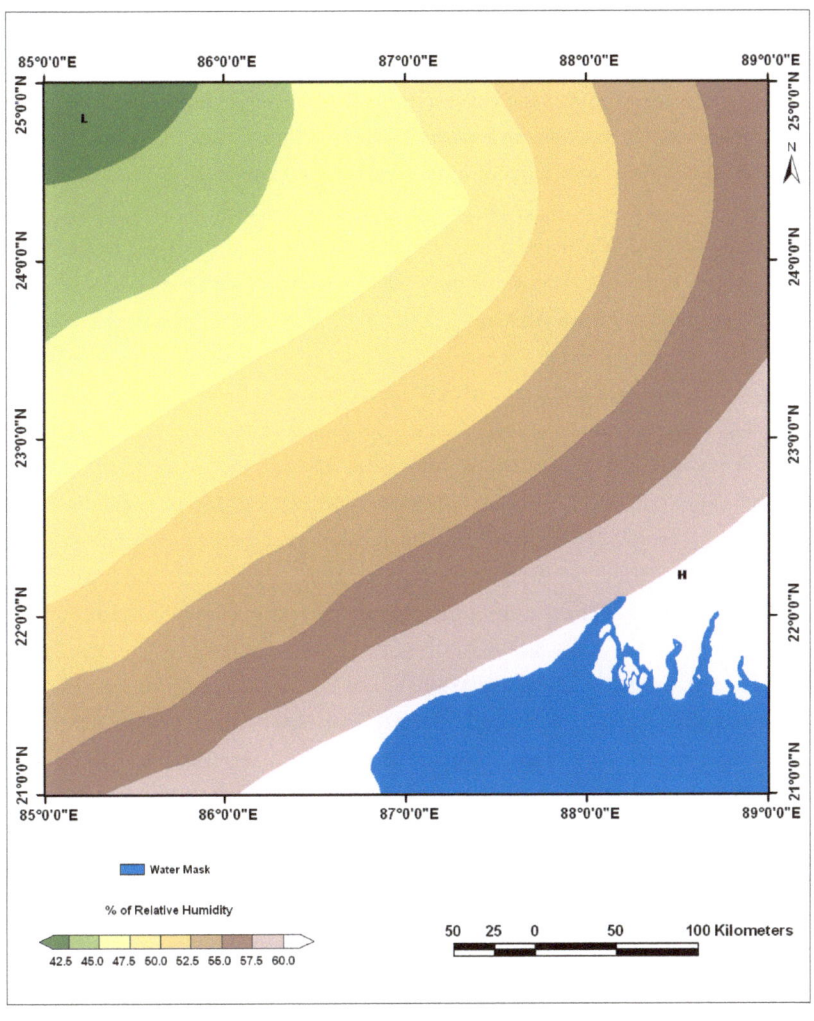

Figure 6.5. Interpolation map of average relative humidity of March (30 years average)

6.5 Climatological modelling

The spatial and temporal distribution of temperature and rainfall has a major impact on the agriculture, environment and economy of many countries. Spatially distributed estimates of rainfall are required inputs to many environmental models. There is also an increasing demand for digital GIS-compatible maps of hydro-meteorological data (Daly et al. 2002). Hydro-meteorological data such as air temperature, humidity and rainfall are typically recorded at the ground level at a limited number of weather stations scattered over a region. Consequently, values have to be estimated at intermediate un-sampled locations in order to generate continuous surfaces for a whole area.

In contrast to the methods described above, artificial neural networks (ANN) are particularly adept at handling massive amounts of data, dealing with complex nonlinear relationships, coping with non-normal and inter-correlated inputs, and allowing the incorporation of additional data and expert knowledge about a particular geographical domain within the estimation process (Bishop 1995). ANNs are also able to select those values from a data set that have a large influence on the outcome, whilst giving little weight to other data. This robustness to redundant data could be used to incorporate flexibility about the number of stations and the number of covariate data sets that should be used on different days to form the best estimate of the rainfall in a local area. Where it is appropriate, other types of data can also be incorporated into the ANN. For example, imagery from rainfall radar could be used to improve estimation of areas where there is evidence that it is, or is not, raining at a given time. ANNs can be used to estimate daily temperature surfaces with similar accuracy to the most sophisticated of the conventional methods, whilst requiring less stringent assumptions to be met about the data conditions (Rigol et al. 2001).

6.6 Modelling of temperature

Surfaces of daily temperature data are relatively continuous and can often be estimated relatively accurately using data from only a small number of nearby stations and using a small number of environmental factors as covariates. For example, after the inclusion of elevation as one covariate, including further covariates yielded only minor further overall improvements in the accuracy of estimations, although additional factors such as distance from the coast, distance from urban areas and local shelter did improve estimations locally for certain types of weather system (Jarvis and Stuart 2001). Altitude/elevation, land use/land cover, soil texture are used for modelling of temperature. The monthly, season and annual average mean temperature is predicted in this modelling. The neural network approach is used to model the monthly average means temperature. For this purpose hundreds of sample points are used as reference point. Existing climate data set of CRU is used as reference data set. Four equations (Equation 1 to 4) are generated as the processes of temperature modelling are discussed in the following paragraph.

$$L_\beta = \frac{\sum_{x,i} \frac{t \cdot x}{M}}{N} \quad \ldots\ldots\ldots\ldots \text{ (Equation 1)}$$

$$S_\gamma = \frac{\sum_{y,i} \frac{t \cdot y}{M}}{N} \quad \ldots\ldots\ldots\ldots \text{ (Equation 2)}$$

$$E_\delta = \frac{\sum_{z,i} \frac{t \cdot z}{M}}{N} \quad \ldots\ldots\ldots\ldots \text{ (Equation 3)}$$

$$L_\beta + S_\gamma + E_\delta = T \quad \ldots\ldots\ldots \text{ (Equation 4)}$$

Where:

$\beta \propto x, i; \gamma \propto y, i; \delta \propto z, i$

t= Observed temperature

x= Weight of land use/ land cover (33% to 40%)

y= Weight of soil texture (12% to 20%)

z= Weight of elevation (30% to 40%)

i= Month, January to December (1 – 12)

M= x + y + z

N= No of treatment sample point

T= Predicted temperature

L_β= Predicted temperature weight by land use

S_γ= Predicted temperature weight by soil

E_δ= Predicted temperature weight by elevation

6.7 Modelling of rainfall

In contrast, estimation of the pattern and amount of rainfall is a more complex task due to its discontinuous and quasi-stochastic nature in space and time. The added complexity of rainfall fields suggests that more factors may need to be included when data are interpolated in order to achieve acceptable accuracy. Altitude/elevation, land use/land cover, relative humidity and cloud cover are used for modelling of rainfall. The monthly total rainfall is predicted in this modelling. Here also neural network approach is used to model the monthly total rainfall. For this purpose hundreds of sample points are used as reference point. Existing climate data set of CRU is used as reference data set. Five equations (Equation 5 to 9) are generated as the processes of rainfall modelling and are being discussed in the following paragraph.

$$E_\alpha = \frac{\sum_{a,i} \frac{r_i a}{f}}{N} \quad \ldots\ldots\ldots\ldots\text{(Equation 5)}$$

$$L_\beta = \frac{\sum_{b,i} \frac{r_i b}{f}}{N} \quad \ldots\ldots\ldots\ldots\text{(Equation 6)}$$

$$H_\gamma = \frac{\sum_{c,i} \frac{r_i c}{f}}{N} \quad \ldots\ldots\ldots\ldots\text{(Equation 7)}$$

$$C_\delta = \frac{\sum_{d,i} \frac{r_i d}{f}}{N} \quad \ldots\ldots\ldots\ldots\text{(Equation 8)}$$

$$E_a + L_b + R_c + C_d = R \quad \ldots\ldots\text{(Equation 9)}$$

Where:

$\alpha \propto a, i$; $\beta \propto b, i$; $\gamma \propto c, i$; $\delta \propto d, i$

a= Weight of elevation (1% to 5%)

b= Weight of land use (1% to 5%)

c= Weight of relative humidity (2% to 35%)

d= Weight of cloud cover (5% to 55%)

f= a + b + c + d

N= Number of reference point

r= Observed rainfall

i = Month; January to December (1 - 12)

R= Predicted rainfall

E_α= Predicted rainfall weight by elevation

L_β= Predicted rainfall weight by land use

H_γ= Predicted rainfall weight by relative humidity

C_δ= Predicted rainfall weight by cloud cover

6.8 Method description of temperature and rainfall modelling

To model monthly, seasonal, annual average temperature and total rainfall of the study area different physical climatic parameters, like land use, soil texture & elevation to predict the temperature and land use, elevation, relative humidity & cloud cover are used for rainfall modelling.

6.8.1 Weight selection for the climatic parameters

We considered cent percent (100%) influence of those parameters for the prediction of monthly average temperature. We selected different weight values for different parameters and for each category of data set in model. Taking the example of land use, there are different types of land use/ land cover classes and as such separate weight values are assigned for each and every class.

A model calculation has been made to choose different weight for the parameters. First we chose the average temperature data set of January. Secondly we found out two points with similar characteristics except the elevation, as for example their relative height difference is 800 m and other characteristics like land use and soil texture are identical. The temperature is measured for both the places and we observed that they are not similar. Then we measured the difference of temperature emanating from the effect of elevation. We made numbers of measurement to find out the average weight value against elevation and it is almost forty percent (40%) of the average mean temperature.

We made this calculation in point A and point B, which had been located at 86.33° E / 21.67° N and 87.17° E / 22.50° N. Soil texture type is 'sandy clay loam' and the land use type is 'forest' in both the points. The elevation of point A and B is 900 m and 100 m. The average mean annual temperature in point A was 24.7° C and B was 26.5° C. The total range of annual average mean temperature according to CRU 30 year's data set was 4.5° C and the temperature range of those points was 1.8° C, which was 40% of the total range. The same calculation process has done to choose the weight value for land use and soil texture. We considered influence of soil texture as twenty percent (20%) while the influence of land use as forty percent (40%). The detailed weight values are tabulated in the Table 6.8 and Table 6.9 as followings-

Table 6.8. Weight values for different parameters for temperature model

Elevation in m	Percent Weight	Land Use/land cover	Percent Weight	Soil Texture Class	Percent Weight
0-100	40	Urban and Built up	40	Loamy Sand	20
100-200	39	Marshy Land	38	Sandy Loam	19
200-300	38	Crop	36	Silty Loam	18
300-400	37	Shrub	34	Loamy	17
400-500	36	Forest	33	Sandy Clay Loam	16
500-600	35	-	-	Silty Clay Loam	15
600-700	34	-	-	Clay Loam	14
700-800	33	-	-	Sandy Clay	13
800-900	32	-	-	Silty Clay	12
900-1000	31	-	-	-	-
>1000	30	-	-	-	-

Table 6.9. Weight values for different parameters for rainfall model

% Cloud Cover	Percent Weight	% Relative Humidity	Percent Weight	Land Use/ land cover	Percent Weight	Elevation in meter	Percent Weight
> 80	55	>80	35	Forest	5	>1000	5
75-80	50	75-80	30	Shrub	4	750-1000	4
70-75	45	70-75	25	Crop	3	500-750	3
65-70	40	65-70	20	Marshy Land	2	250-500	2
60-65	35	60-65	15	Urban and Built up	1	<250	1
55-60	30	55-60	10	-	-	-	-
50-55	25	50-55	7.5	-	-	-	-
40-50	15	40-50	5	-	-	-	-
30-40	10	0-40	2.5	-	-	-	-
0-30	5	-	-	-	-	-	-

6.8.2 Temperature and rainfall according to selected parameters

To calculate the temperature and rainfall at first we selected hundreds of sample point with known value or observed value of temperature and rainfall. The observed temperature or rainfall is considered for each category of the different parameters according the weight value. Then we found out the monthly average temperature and total rainfall for each category of parameters according to the hundreds (100) of sample points characteristics. Next we calculated the temperature and rainfall for all parameters and their categories (Appendix 9 & 10). Finally the result is distributed to all other location according to different parameter in the study area.

6.9 GIS methodology for the modelling

Using ERDAS IMAGINE and ArcGIS software different raster layers are generated for each parameter used for temperature and rainfall modelling. SRTM data is used to generate DEM for the study area. There are some data gaps in the SRTM data set. We had to use topographical maps in order to fill the data gaps. Using the 3-D analysis process the digital elevation model is generated from contours, which had been digitized from topographical maps. Finally, the elevation map had been generated. The land use /land cover data set is generated from the digital image classification of LANDSAT, ETM+ satellite images. Soil texture data set is generated by the process of digitization from soil region maps of the area. West Bengal soils sheet of national bureau of soil survey (NBSS), soil region map of national atlas & thematic mapping organization are used for this purpose. The final model is created by the model maker in the GIS environment using simple equations. Land use/land cover, soil texture and digital elevation model are used as input parameters for temperature modelling and land use/land cover, elevation, cloud and humidity for rainfall modelling. In the output, we developed new raster layers with the predicted temperature and rainfall value. The same procedure is extended for all the season and annual temperature and rainfall modelling. The schematic models of temperature and rainfall are shown in the Figure 6.6. The model outputs (temperature and rainfall) for the month of June are displayed in Figures 6.7 and 6.8. Total rainfall of premonsoon season is displayed in Figure 6.9.

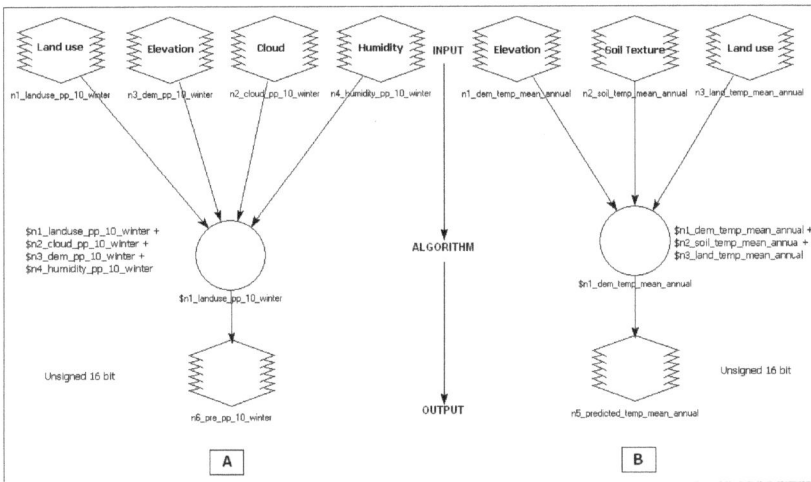

Figure 6.6. Schematic models of (A) rainfall and (B) temperature

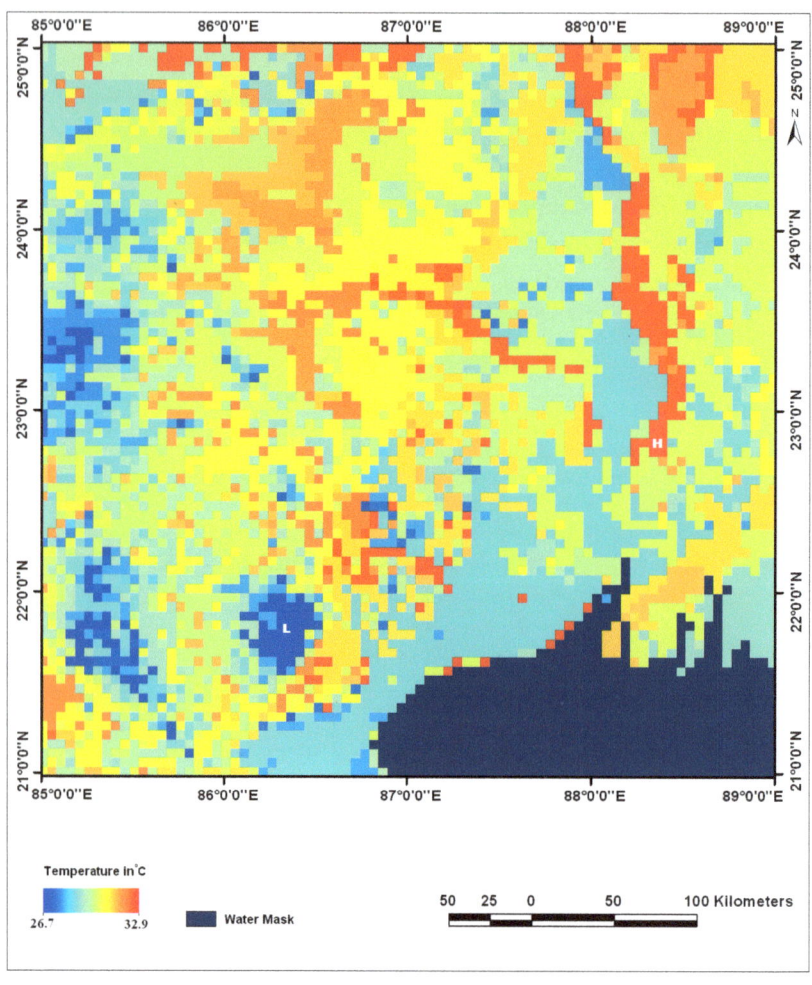

Figure 6.7. Predicted average mean temperature for the month of June based on temperature modelling

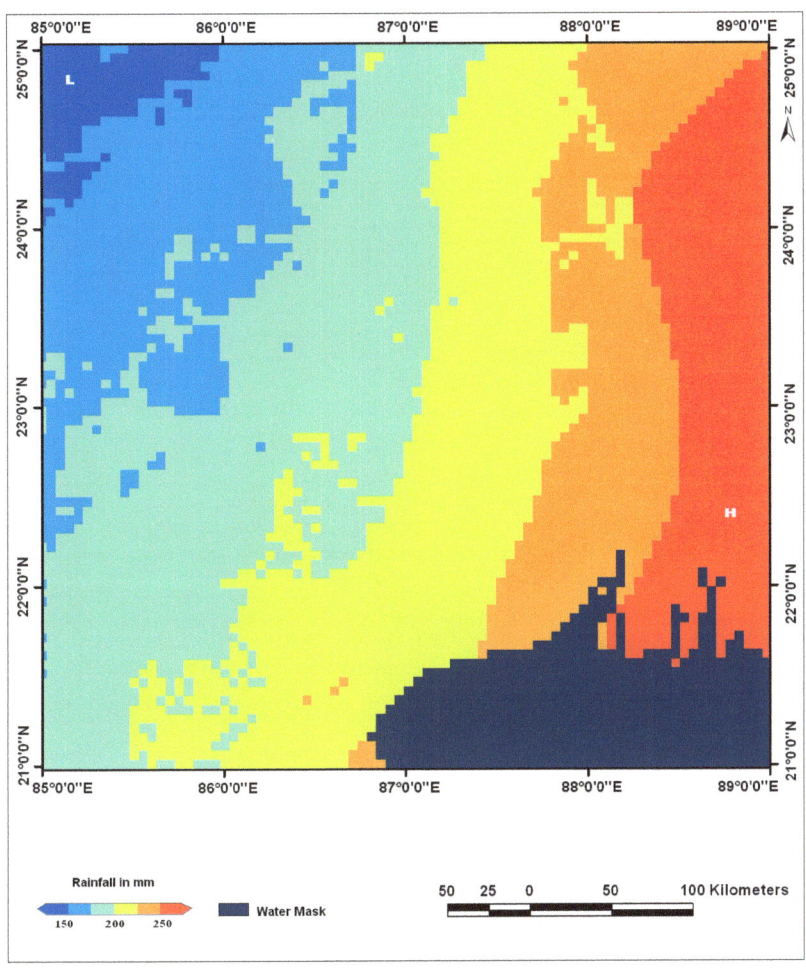

Figure 6.8. Predicted total rainfall for the month of June based on rainfall modelling

6.10 Methodological flow chart for modelling of temperature and rainfall

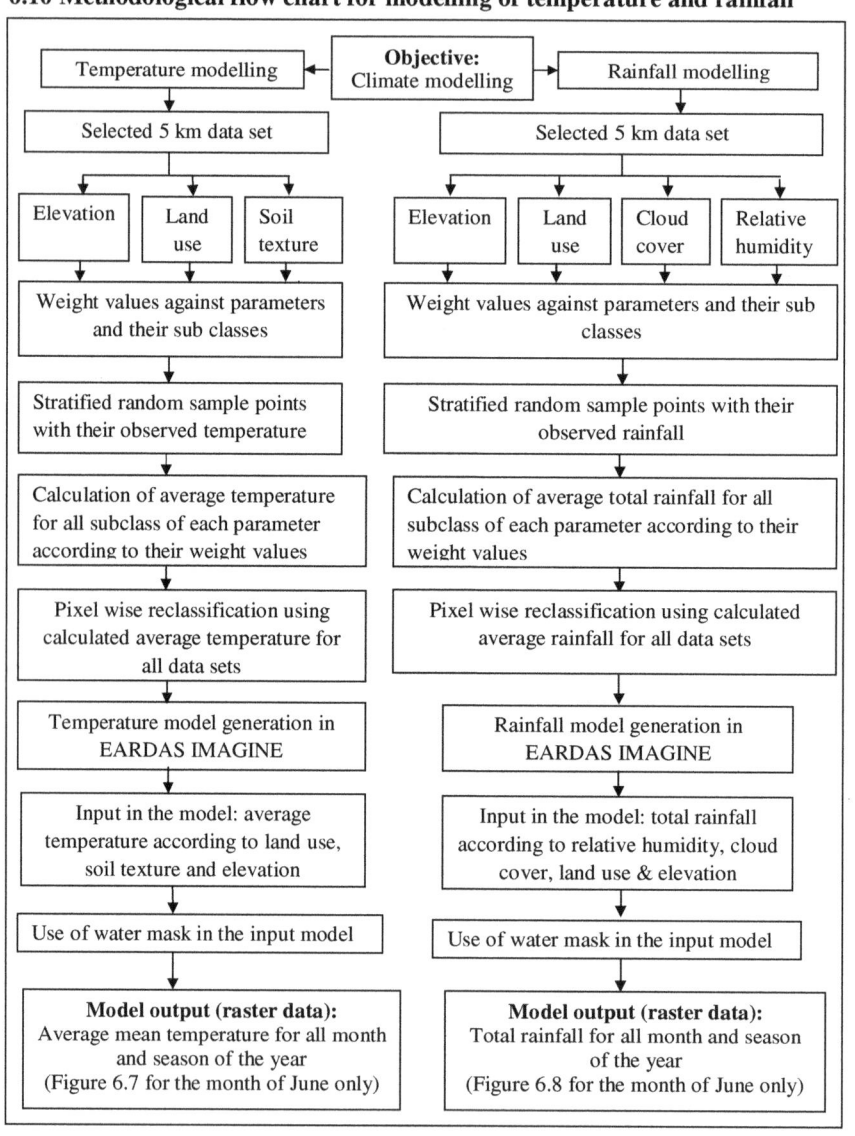

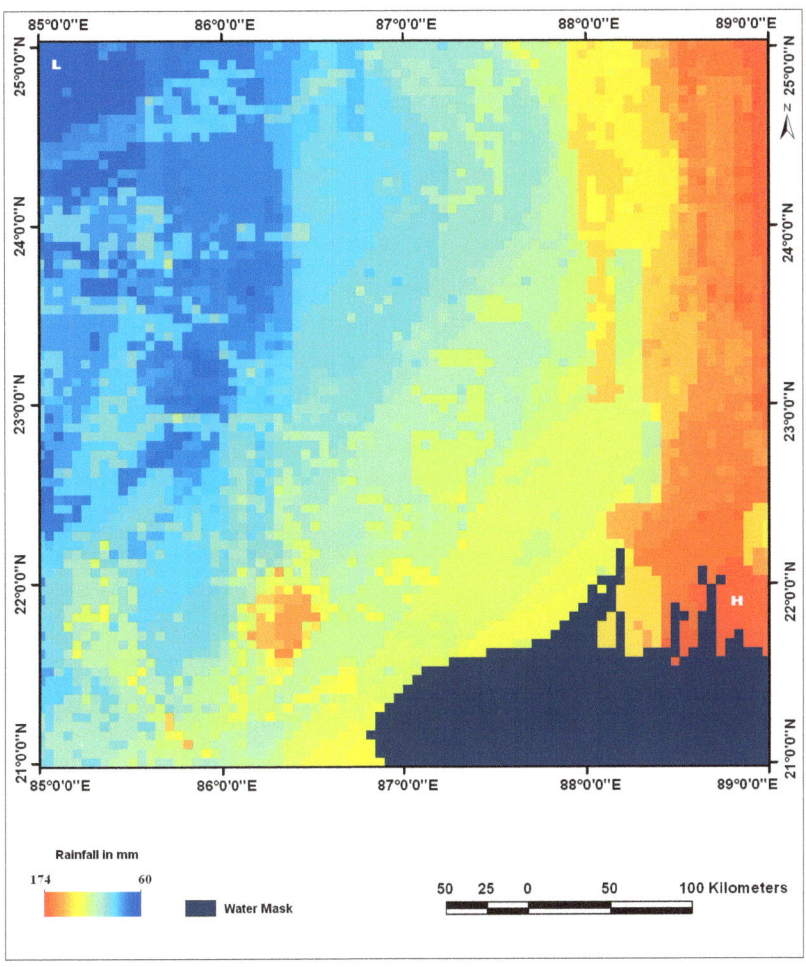

Figure 6.9. Predicted total rainfall for premonsoon season based on rainfall modelling

6.11 Results and verification analysis

6.11.1 Temperature modelling

Standard deviation errors are calculated for monthly mean temperature, seasonal mean temperature and annual mean temperature and are displayed in Table 6.10. It reveals that predicted temperature over 85% of the study has standard deviation error less than 1.0, irrespective of the month or season. Hence it justifies the applicability of the procedure used for predicting mean temperature. The standard deviation value is less in monsoon season while it is relatively high in the winter months and as a whole in the winter season over the entire study area. Premonsoon and postmonsoon seasons come in between.

Table 6.10. Standard deviation error estimation of predicted mean temperature using CRU-30 year's data set

Month	% of area coverage against Standard deviation error			
	0 - 0.5	0.5- 1.0	1.0 - 1.5	1.5 - 2.0
January	55.26	32.56	10.58	1.60
February	52.06	32.49	13.37	2.08
March	61.56	31.32	6.84	0.27
April	76.41	22.27	1.25	0.07
May	70.65	22.75	6.27	0.33
June	77.96	19.33	2.65	0.07
July	80.38	18.73	0.89	0.0
August	76.96	21.77	1.25	0.02
September	74.07	24.78	1.12	0.03
October	72.80	25.46	1.70	0.03
November	64.74	31.01	4.11	0.14
December	62.66	31.03	6.22	0.09
Average Annual	72.99	25.70	1.22	0.09
Pre-Monsoon	70.31	26.84	2.61	0.24
Monsoon	81.07	18.23	0.70	0
Post-Monsoon	69.18	28.49	2.29	0.05
Winter	57.10	32.58	9.59	0.74

An attempt has also been made here to compare the prediction with the observed values at different weather stations, maintained by IMD. Twenty such weather stations have been selected randomly according to availability of the data. These stations are located in West Bengal, Jharkhand and Orissa (Figure 6.10). They are Digha, Contai, Midnapur, Bankura, Krishnanagar, Purilia, Kolkata, Diamondharbour, Dumdum, Durgapur, Uluberia, Burdwan, Santiniketan, Baripada, Haldia, Balasore, Ranchi, Jamshedpur, Chaibasa and Hazaribag.

Figure 6.10. Location map of different weather stations maintained by IMD over the study area

Table 6.11. Estimated vs. observed temperature

Monthly mean Temperature (° C)	March	April	May	Monthly Temperature (° C)	March	April	May
Station-1	Digha			Station-11	Diamond harbour		
Observed	26.1	28.4	29.2	Observed	27.2	29.5	30.0
Predicted	25.5	28.9	30.2	Predicted	26.2	29.9	31.2
Station-2	Balasore			Station-12	Contai		
Observed	27.3	29.8	30.5	Observed	26.9	29.0	29.5
Predicted	25.5	28.9	30.2	Predicted	25.5	28.9	30.2
Station-3	Midnapur			Station-13	Dumdum		
Observed	27.8	31.2	31.4	Observed	27.1	30.1	30.4
Predicted	26.5	30.3	31.8	Predicted	26.8	30.0	31.1
Station-4	Ranchi			Station-14	Durgapur		
Observed	24.2	28.1	30.1	Observed	23.4	30.85	31.2
Predicted	24.6	28.5	30.0	Predicted	26.2	29.9	31.2
Station-5	Bankura			Station-15	Uluberia		
Observed	29.0	29.7	32.8	Observed	26.9	29.8	30.2
Predicted	27.8	31.4	32.7	Predicted	26.2	29.9	31.2
Station-6	Jamshedpur			Station-16	Burdwan		
Observed	27.3	30.9	32.7	Observed	27.5	31.2	32.0
Predicted	27.5	31.3	32.5	Predicted	27.8	31.4	32.7
Station-7	Chaibasa			Station-17	Santiniketan		
Observed	26.4	30.4	31.6	Observed	26.3	30.3	30.1
Predicted	26.1	30.1	31.5	Predicted	26.8	30.0	31.1
Station-8	Krishnanagar			Station-18	Baripada		
Observed	25.7	29.7	30.0	Observed	27.5	31.1	31.1
Predicted	26.2	29.9	31.2	Predicted	26.5	30.3	31.8
Station-9	Purulia			Station-19	Haldia		
Observed	25.6	29.7	30.9	Observed	27.4	29.3	29.7
Predicted	26.1	30.1	31.5	Predicted	26.0	29.3	31.0
Station-10	Kolkata			Station-20	Hazaribag		
Observed	27.6	30.3	30.8	Observed	23.3	27.9	29.9
Predicted	26.8	30.0	31.1	Predicted	24.6	28.5	30.0

Comparison has been made only for the premonsoon months due to the availability of the data (Table 6.11). It reveals that the difference in mean monthly temperature is within 1 degree C for almost all stations irrespective of the months in premonsoon season.

6.11.2 Rainfall modelling

Standard deviation errors are calculated for monthly total rainfall and seasonal total rainfall as displayed in Table 6.12. It reveals that the predicted rainfall over a large part of the area has standard deviation error less than 10, irrespective of the month or season, justifying the applicability of the procedure used for predicting total rainfall. However, standard deviation value is more in the monsoon region compared to the other seasons.

Table 6.12. Standard deviation error estimation of predicted total rainfall using CRU-30 year's data set

Month	% of area coverage against Standard deviation error			
	0 - 10	10 - 20	20 - 30	30 - 40
January	100.0	0.0	0.0	0.0
February	93.0	7.0	0.0	0.0
March	98.3	1.7	0.0	0.0
April	75.7	24.3	0.0	0.0
May	71.2	26.8	2.0	0.0
June	54.3	38.1	4.3	3.3
July	61.0	25.7	8.5	4.8
August	53.3	28.7	13.4	4.6
September	55.4	26.8	13.4	4.4
October	62.5	25.8	8.5	3.1
November	88.5	11.5	0.0	0.0
December	100.0	0.0	0.0	0.0
Pre-Monsoon	80.7	17.6	1.7	0.0
Monsoon	57.0	29.8	8.9	4.3
Post-Monsoon	75.1	18.7	5.3	0.9
Winter	96.7	3.3	0.0	0.0

Similar comparison like temperature has also been attempted here to compare the predicted rainfall with the observed values at different weather stations, maintained by IMD. Results are shown in the Table 6.13. It reveals that the predicted values are close to the observed values though there are stations where on some occasions prediction is poor.

Table 6.13. Estimated vs. observed total rainfall

Monthly Rainfall (in mm)	March	April	May	Monthly Rainfall (in mm)	March	April	May
Station-1	Digha			Station-11	Diamond harbour		
Observed	31.8	30.5	97.1	Observed	27.3	49.9	108.7
Predicted	23.3	38.6	63.0	Predicted	23.7	41.1	68.3
Station-2	Balasore			Station-12	Contai		

Observed	51.7	80.9	127.8	Observed	34.4	41.7	74.3
Predicted	21.7	34.7	63.0	Predicted	23.3	38.6	63.0
Station-3	*Midnapur*			*Station-13*	*Dumdum*		
Observed	52.4	68.4	139.4	Observed	22.9	45.8	113.3
Predicted	23.7	37.2	62.6	Predicted	20.9	38.6	90.1
Station-4	*Ranchi*			*Station-14*	*Asansol*		
Observed	21.9	21.2	60.8	Observed	27.3	22.0	78.0
Predicted	17.8	20.3	48.9	Predicted	13.8	23.9	51.6
Station-5	*Bankura*			*Station-15*	*Uluberia*		
Observed	35.9	38.8	88.4	Observed	19.8	42.9	85.0
Predicted	21.3	33.2	62.6	Predicted	20.9	38.6	81.9
Station-6	*Jamshedpur*			*Station-16*	*Burdwan*		
Observed	28.9	28.1	67.1	Observed	39.4	42.7	85.4
Predicted	21.3	29.3	54.3	Predicted	19.3	35.2	65.1
Station-7	*Chaibasa*			*Station-17*	*Santiniketan*		
Observed	20.5	31.7	69.8	Observed	38.0	44.0	90.5
Predicted	18.1	22.0	51.1	Predicted	19.8	31.2	59.4
Station-8	*Krishnanagar*			*Station-18*	*Baripada*		
Observed	19.2	44.6	62.6	Observed	33.5	38.7	98.9
Predicted	20.9	34.7	81.9	Predicted	23.7	37.2	62.8
Station-9	*Puruilia*			*Station-19*	*Haldia*		
Observed	29.1	35.9	70.4	Observed	27.3	49.9	95.5
Predicted	17.2	27.7	51.7	Predicted	26.1	41.1	76.5
Station-10	*Alipur*			*Station-20*	*Hazaribag*		
Observed	22.9	49.3	103.7	Observed	17.7	16.1	53.4
Predicted	20.9	38.6	81.9	Predicted	17.8	20.3	45.3

CHAPTER 7

MODELLING OF TEMPERATURE AND RAINFALL IN WEST BENGAL, INDIA

CHAPTER OUTLINE

- Introduction to the Study area
- Digital Data set of Input parameters
- Results and Discussion

7.1 Introduction to the Study area

The study area is West Bengal in the eastern India. The study area corresponds to the LANDSAT-7 ETM+ imageries with path row numbers 138_43, 138_44, 138_45, 139_43, 139_44, 139_45, 140_43, 140_44 and 140_45. The region comprises with a semi-arid part in the west and south-west and relatively more moist and cool in the south and the east. Being the tail end of Chottonagpur plateau, the terrain of the semi-arid part is more complex with higher elevation and permanent vegetation compared to the south and the eastern part of the study area. The region experiences four seasons having typical weather systems: monsoon (June through September) season, characterized by south-west monsoonal flow having a monsoon trough line through the region causing most of the annual rainfall; post-monsoon season (October through November), being the transition season between the monsoonal flow and westerlies of winter season (December through February) in the upper levels, region experiences tropical storm on occasions; a part of the region on the other hand experiences the impact of western disturbances in winter; and frequent thunderstorm activity in the premonsoon season (March through May). Therefore, throughout the year the study region experiences synoptic scale disturbances, such as monsoons and mesoscale weather systems leading to more complex patterns of climatic variables. This emphasizes the need for a better procedure to map the climatic variables in absence of dense network of observations.

7.2 Digital Data set of Input parameters

7.1.1Land use/land Cover database

The resulted land use/ land cover map from the satellite image (Figure 1) reveals that, in general, the western and northern parts of the study area are relatively more heterogeneous having forested area, crop land and shrub land etc. and therefore the climatic variables patterns are likely to be more complex. While the south and south-east parts of the study area are less heterogeneous except the more complex coastal areas.

The study area contains forest (~4%) which is mainly concentrated in the western and northern parts of the study area. About 12% of the study area is covered by shrub. Agriculture land takes the lead covering about 79% of the land while less than 1% land is used for residential and industrial purposes. 'Water' class (4%), mostly belongs to the sea, is concentrated in the southern part of the study area (Table 1).

Table 1. Land use/ land cover class statistics of West Bengal

Sl. No.	Land use/ land cover	Histogram (No of pixel)	Area in sq km	% of area cover
1	Water	4369	3538.89	4.01
2	Forest/vegetation cover	4192	3395.52	3.84
3	Shrub land	13167	10665.27	12.07
4	Fallow/grass/single cropped land	35904	29082.24	32.92
5	Agriculture land	50098	40579.38	45.93
6	Urban and build up	1336	1082.16	1.22
	Total	109066	88343.46	100.00

7.1.2 Soil texture database

Using NBSS soil sheets the soil texture database of the study area has been generated with the help standard GIS methodology. All the soils are categorized into nine (9) soil texture classes according to their characteristics, namely loamy sand, sandy loam, silty loam, sandy clay loam, clay loam, silty clay loam, clay loam, sandy clay, and silty clay. Sandy clay loam takes the maximum (~26% of the area) share followed by silty clay loam (~24% of the area). Therefore these two soil textures contribute about half of the area (Figure 2 and table 2).

Table 2. Soil texture class statistics of West Bengal

Sl. No.	Soil texture classes name	Histogram (No of pixel)	Area (in sq km)	% of area
1	Water	1851	1499.31	1.70
2	Loamy sand	18493	14979.33	16.96
3	Sandy loam	6315	5115.15	5.79
4	Silty loam	750	607.5	0.69
5	Loamy	3111	2519.91	2.85
6	Sandy clay loam	28648	23204.88	26.27
7	Silty clay loam	26060	21108.6	23.89
8	Clay loam	1683	1363.23	1.54
9	Sandy clay	253	204.93	0.23
10	Silty clay	21902	17740.62	20.08
	Total	109066	88343.46	100.00

7.1.2 Topographic database

Shuttle Radar Topography Mission (SRTM) data and topographical maps are used to generate the Digital Elevation Model (DEM) with the help of surfacing techniques in the GIS environment (Figure 3 and table 3). The elevation of the eastern and southern parts of the study area is less than 50 m. The slope of the topography also is very less in the south and eastern part while the north part is relatively more elevated (>3000 m) having higher slope than the other parts. Air temperature is likely to decrease on such locations by about 6.5 degree C considering the normal lapse rate of temperature 0.65° C every 100 meter. Therefore consideration of topographic height in spatial interpolation of climatic variables is one of the essential requirements.

Table 3. Elevation class statistics of West Bengal

Row No.	classes (Elevation in m)	Histogram (No of pixel)	Area (in sq km)	% of area
1	Less Than 50	69144	56006.64	63.40
2	50 - 100	18794	15223.14	17.23
3	100 – 200	11935	9667.35	10.94

4	200 – 400	5872	4756.32	5.38
5	400 – 600	769	622.89	0.71
6	600 –800	418	338.58	0.38
7	800 – 1000	357	289.17	0.33
8	1000 – 2000	1324	1072.44	1.21
9	2000– 3000	387	313.47	0.35
10	More than 3000	66	53.46	0.06
	Total	109066	88343.46	100.00

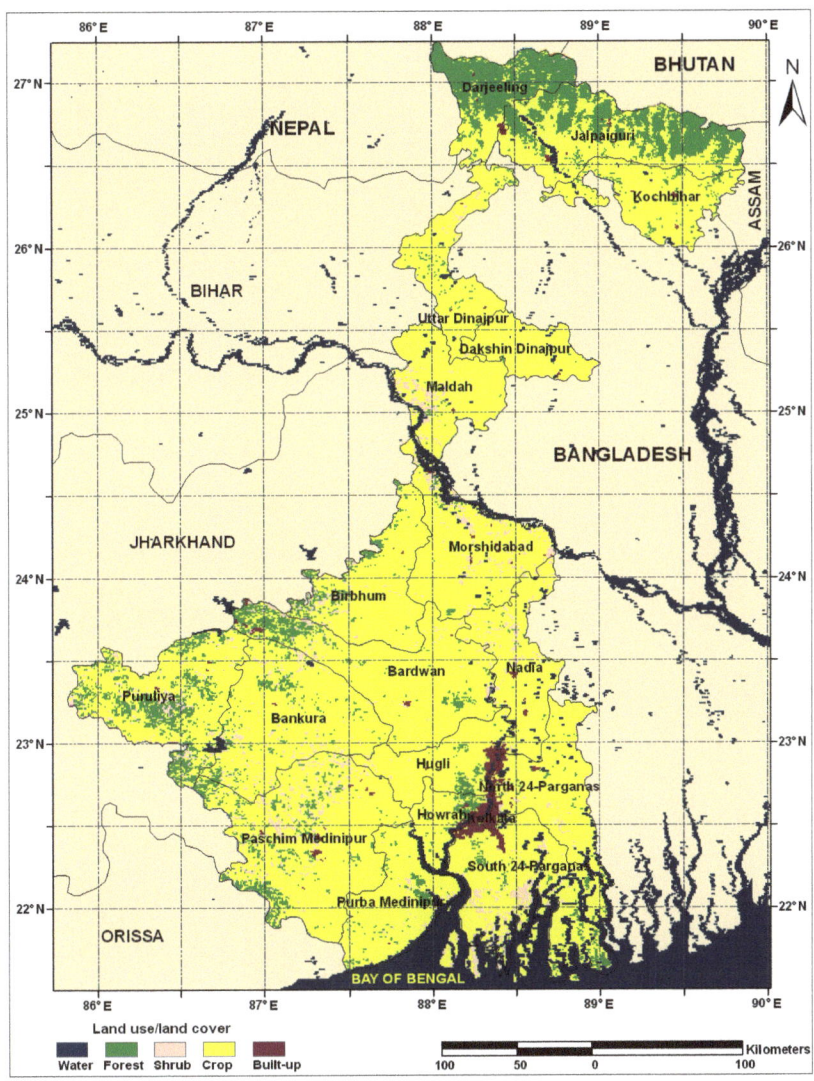

Figure 7.1 Land use/land cover map of West Bengal generated from LANDSAT 7 satellite image

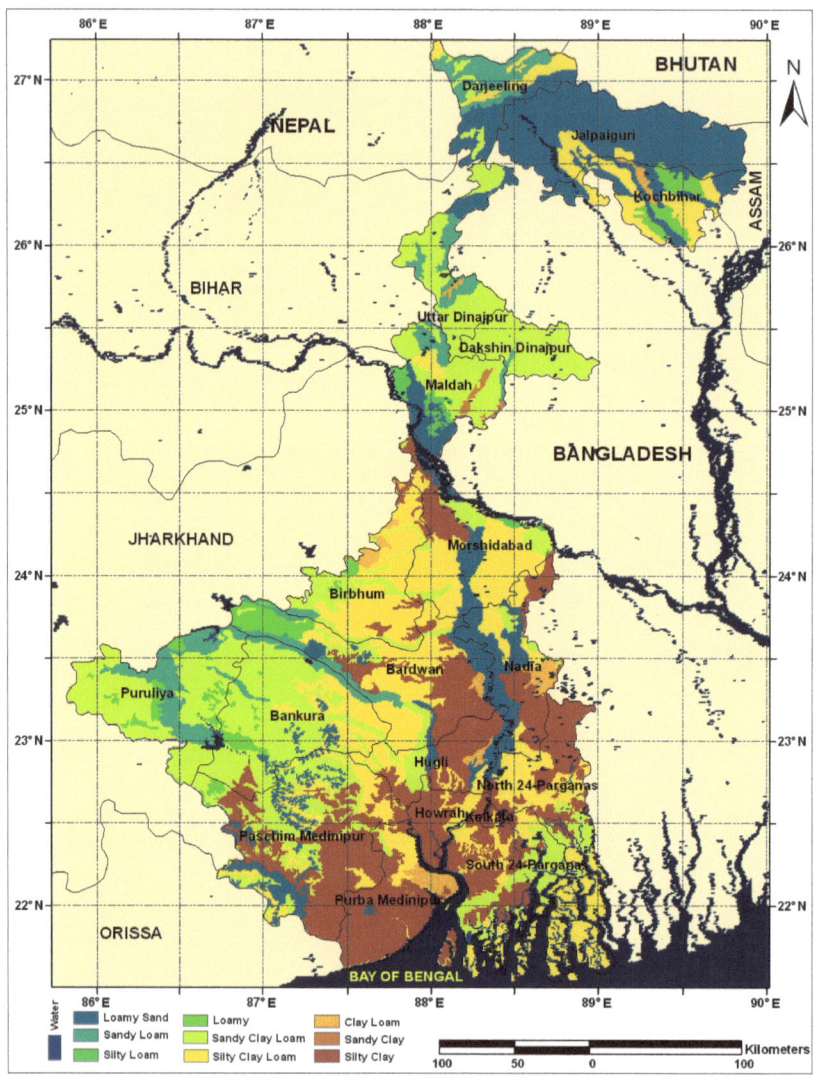

Figure 7.2. Soil texture map of West Bengal generated from NBSS soil sheet

123

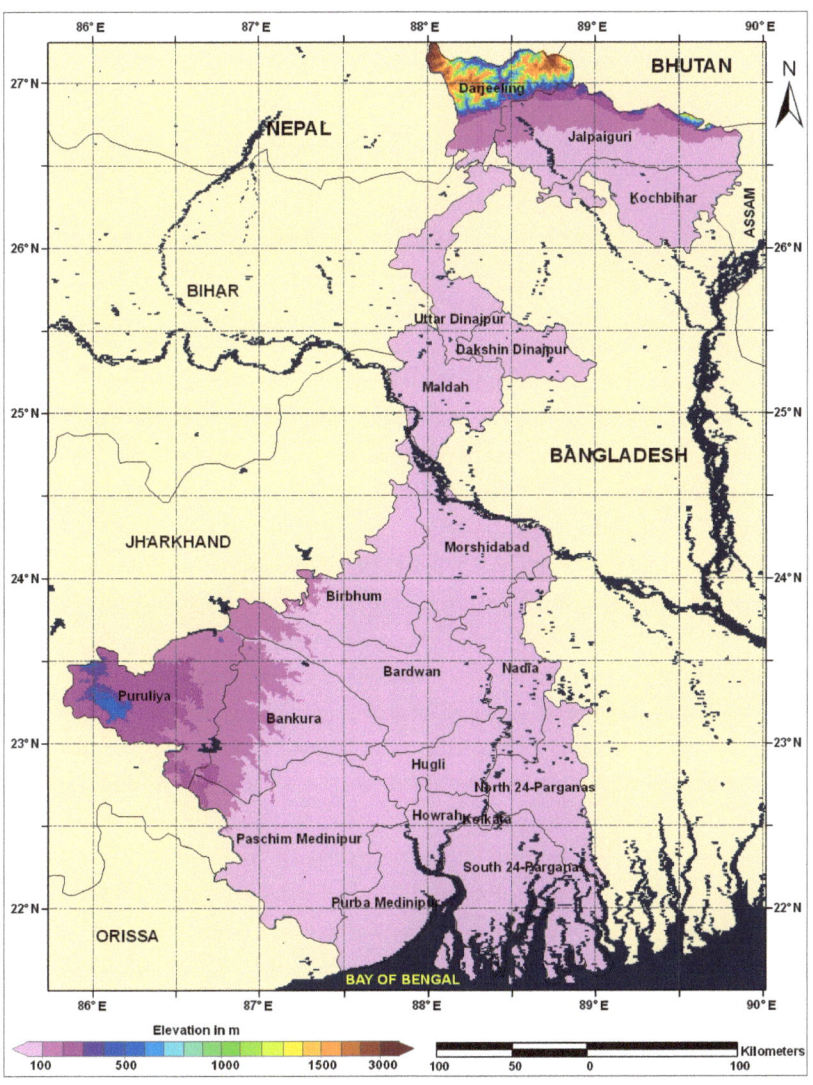

Figure 7.3. Elevation model of West Bengal generated from topographical map and SRTM data

7.23 Results and Discussion

Land use/land cover, soil texture and digital elevation model were used as input parameters for temperature modelling and Land use/land cover, digital elevation model, cloud cover and relative humidity for rainfall modelling. In the output, we developed a new raster layer with the predicted mean temperature and total rainfall. The model outputs (temperature and rainfall) for the winter and pre-monsoon season are displayed in Figures 7.4 and 7.5. Total rainfall of the monsoon season and annual rainfall are displayed in Figure 7.6 and 7.7. The prediction was relatively poor in the winter months, possibly due to lack of consideration of the impact of western disturbances. In winter season, extra-tropical disturbances (ETD) passes through the Himalayan region and while passing through the north of the study area, an induced low pressure system forms resulting sharp changes of temperature. The region also experienced little amount of rainfall. Such a disturbance is unlikely to be considered in the present model. Therefore in order to have a better prediction in winter month, principal wind pattern is to be considered as a parameter. Usually, the impact of ETD, on occasions, extended up to the middle of March. Thus the same weather disturbance is expected to make the performance poor the prediction is better in monsoon season. Being synoptic phenomena, the entire study area gets affected by the system unlike mesoscale disturbances such as tropical storms in post-monsoon season or thunderstorms in pre-monsoon season. Being the transition period from the winter; both the pre-monsoon and post-monsoon seasons are susceptible for tropical storms. Standard deviation errors were calculated for monthly total rainfall and seasonal total rainfall. It revealed that percentage of area coverage was very less having standard deviation errors greater than 10.0, irrespective of the month or seasons. Hence it justified the applicability of the procedure used for predicting total rainfall. However, percentage of area coverage with standard error less than 10.0 was below 75% for monsoon seasons. Thus the prediction was quite poor in the monsoon months, possibly due to lack of consideration of the impact of monsoonal uncertainty. In the summer monsoons the winds carried moisture from the Indian Ocean and bring heavy rains from June to September in this study region. The monsoon accounts for 80% of the rainfall in the study region. The summer monsoon consisted of alternating wet and dry events known as active and break period. During active periods, low pressure systems brought frequent thunderstorms and heavy rain, whereas break period looked typically bright and sunny. The timing and duration of active and break periods accounted for much of the year to year variation in the monsoon, dry years with frequent and long lived break.

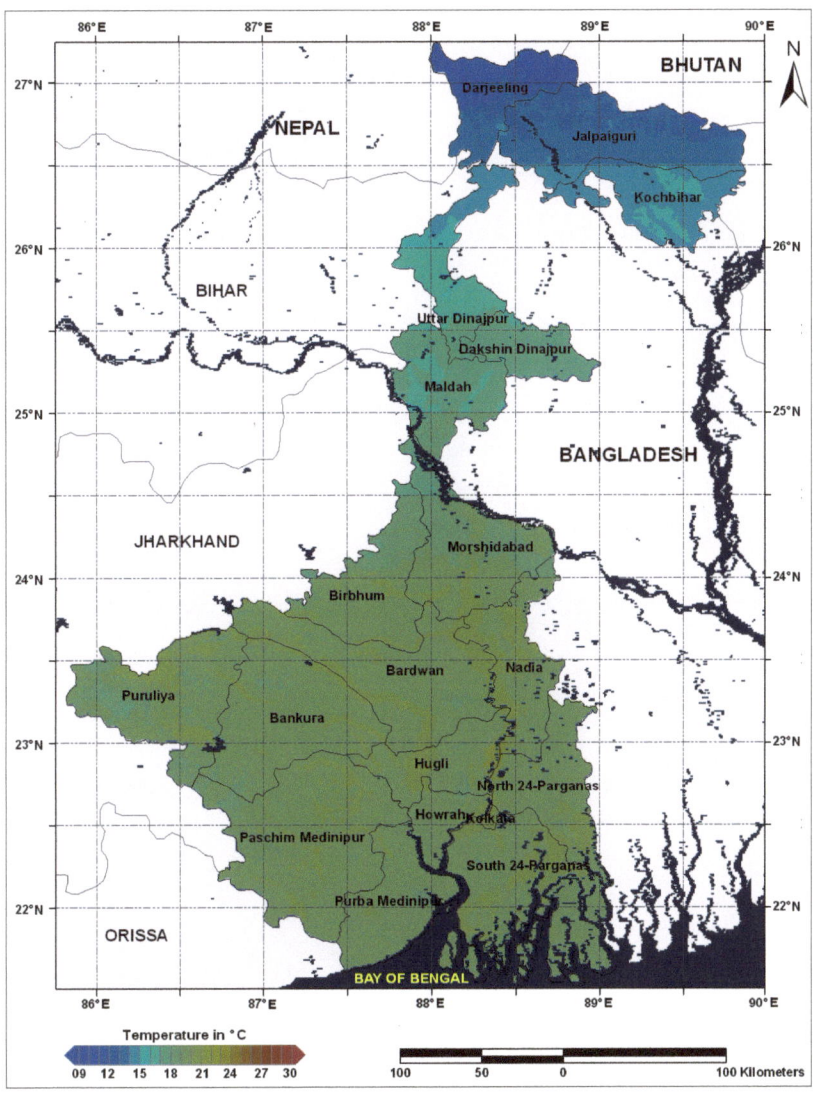

Figure 7.5. Modelled mean temperature of winter season of West Bengal

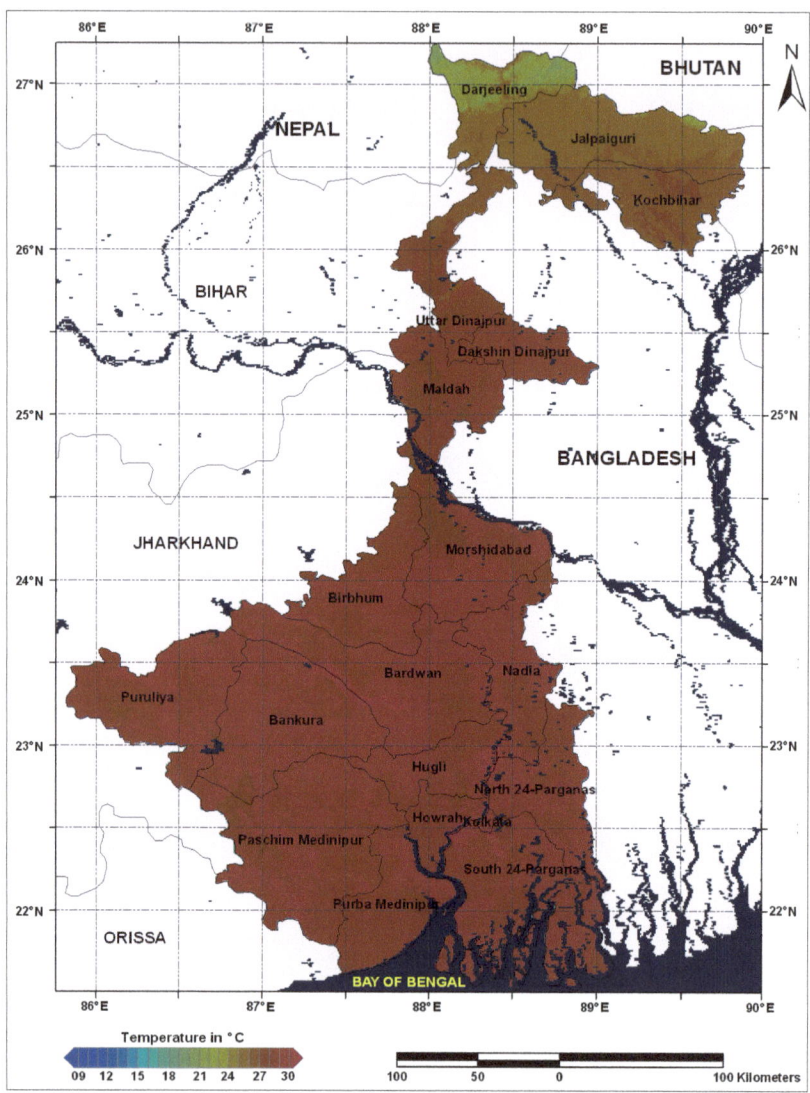

Figure 7.5. Modelled mean temperature of pre-monsoon of West Bengal

127

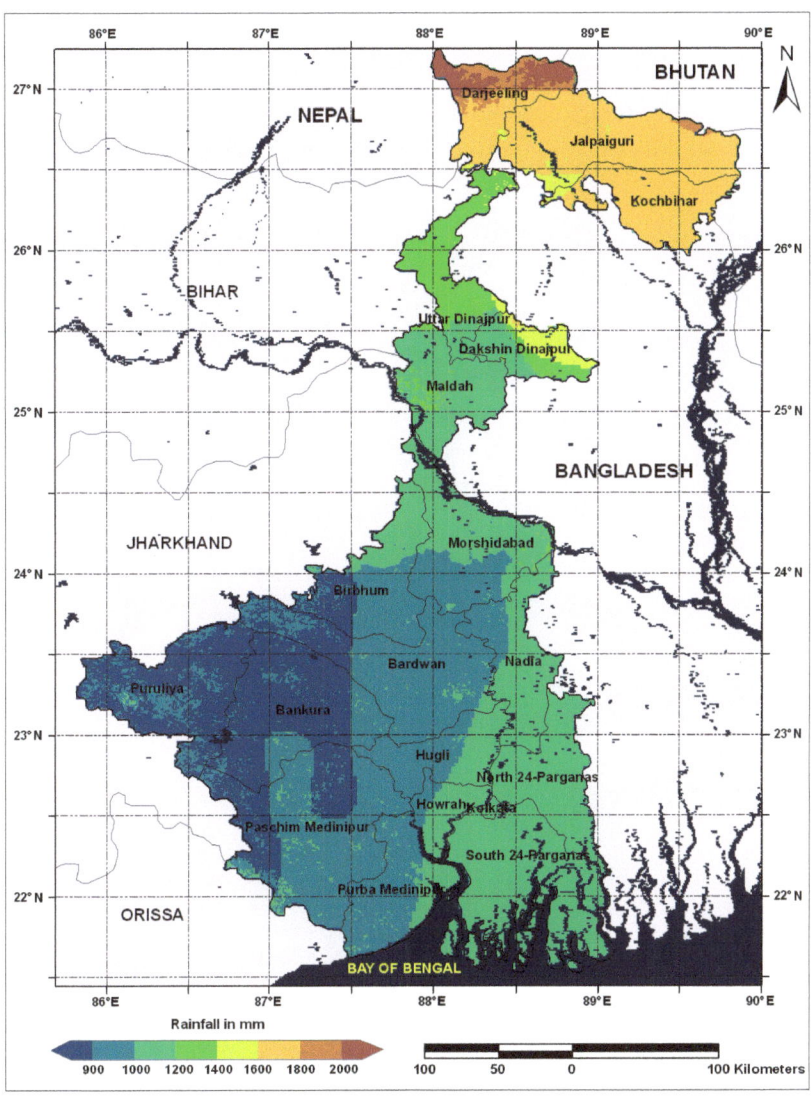

Figure 7.6. Predicted total rainfall in monsoon season of West Bengal

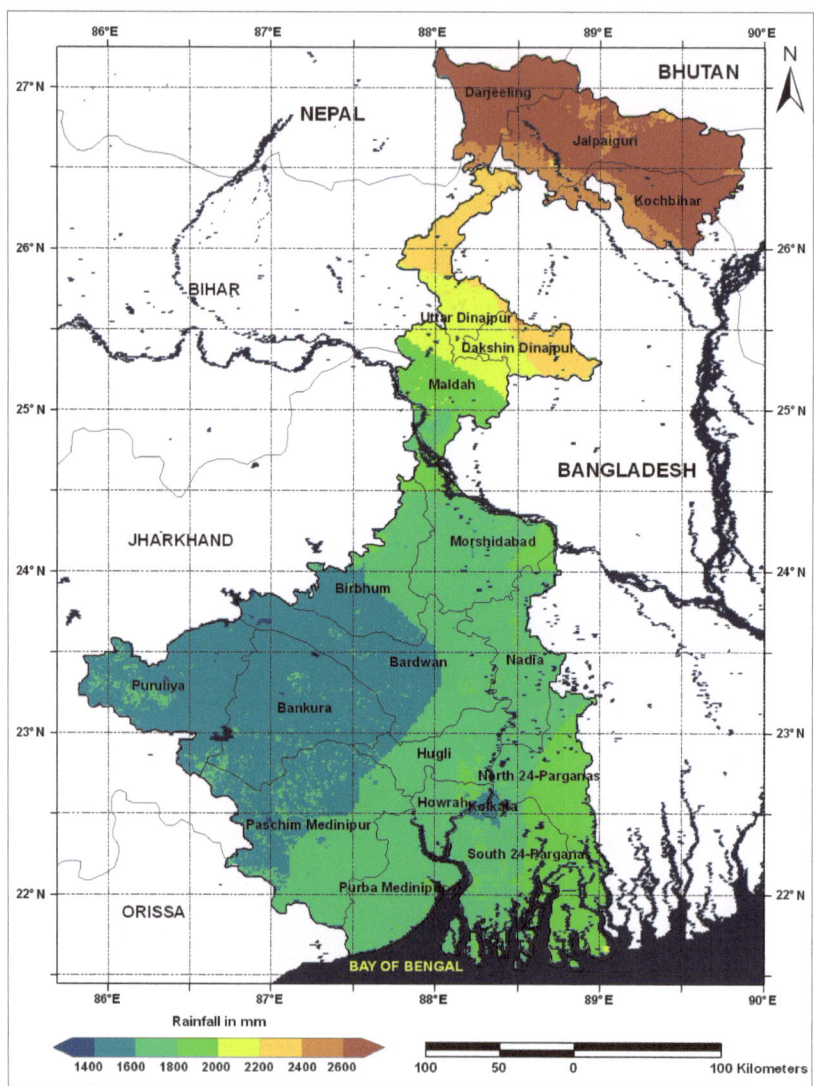

Figure 7.7. Predicted total annual rainfall of West Bengal

CHAPTER 8

DISCUSSION AND CONCLUSIONS

CHAPTER OUTLINE

- Discussion and conclusions
- Recommendations for future studies

8.0 DISCUSSION AND CONCLUSIONS

8.1 Discussion and Conclusions

It is indeed an important task to provide 'wall to wall' reliable climate variable fields that are derived from discrete point observations, on a daily, monthly, or annual timescale basis is an important task. This is particularly more challenging where observations are sparse. Therefore recognizing the necessity of data interpolation, lots of approaches are being tried by the researchers and viability of the approaches is also being tested for different types of requirements. As the values of the climate variables depend on a number of factors, e.g. the topography (height above sea level, orientation of slopes, curvature, etc.), aerodynamic roughness, vegetation coverage, soil types, distance from the sea etc., an obvious choice for spatial interpolation of the climate variables is to seek statistical relationship between the variables and hence predict the climatic variables for locations with known geographical factors. This is being done in climatological modelling using GIS and it has become a very powerful tool in agricultural research and natural resource management. The present work is an attempt to model the climatic variables, temperature and rainfall over a tropical region in the eastern part of India (21° N - 25° N and 85° E - 89° E). The region has limited point observations of the meteorological variables. It also experiences several weather disturbances which pass through the region throughout the year.

The study area has unique geographical location with the Himalayas in the north and the Bay of Bengal in the south and is also endowed with diverse characteristics of landforms such as hills, plateaus and uplands, flat plains and low-lying delta areas. A detailed description of the geographical setting, as given in chapter 2, indicates widespread climatological variations of the study area.

Wetlands are being destroyed due to erosion at the coastal areas and due to huge increasing pressure of the habitants in the inland areas. Moreover land use change is a regular process as a result of deforestation, afforestation, increase in cultivated and built up areas. As the land use/land cover categories greatly alter the climatic characteristics of the area, as has been detailed in Chapter 3. LANDSAT 7, ETM+ satellite imagery are used to determine the land use/ land cover of the study area with the help of pixel by pixel standard classification methodology. Three data sets are generated having resolution of 30 m, 1 km and 5 km. The results reveal that, in general, the western part of the study area is relatively more heterogeneous having forested area, crop land and shrub land etc. and therefore the climatic variables patterns are likely to be more complex. While the eastern part of the study area is less heterogeneous except the more complex coastal areas. The study area contains forest (~12%) which is mainly concentrated in the western part of the study area. About 12% of the study area is covered by shrub. Agriculture land takes the lead covering about 62% of the land while less than 1% land is used for residential and industrial purposes. 'Water' class (11%), mostly belongs to the sea, is concentrated in the southern part while the 'marshy land' (less than 2%) is strewn in the eastern part of the study area. Using NBSS & LUP soil sheets the soil texture database of the study area has been generated with the help standard GIS methodology. All the soils are categorized into nine (9) soil texture classes according to their characteristics, namely loamy sand, sandy loam, silty loam, sandy clay loam, clay loam, silty clay loam, clay loam, sandy clay, and silty clay. Sandy clay loam takes the maximum (~30% of the area) share followed by sandy loam (~20% of the area). Therefore these two soil textures contribute about half of the area.

Shuttle Radar Topography Mission (SRTM) data and topographical maps are used to generate the Digital Elevation Model (DEM) with the help of surfacing techniques in the GIS environment (Chapter 4). Three data sets are generated to represent the digital elevation model having spatial resolution 90 m, 1 km and 5 km respectively. The elevation of the eastern and southern parts of the study area is less than 100 m. The slope of the topography also is very less in the south and eastern part while the south-west part is relatively more elevated (>1000 m) having higher slope than the other parts. Air temperature is likely to decrease on such locations by about 6.5 degree C considering the normal lapse rate of temperature $0.65°$ C every 100 meter. Therefore consideration of topographic height in spatial interpolation of climatic variables is one of the essential requirements.

Using the thermal bands, hot spots are identified through thermal modelling according to the availability of the real time satellite images (Chapter 5). The thermal plant area within the study area (Kolaghat Thermal Plant) and the brick kiln industrial regions show the high temperature value for their thermal influencing properties with an ostensibly high emittance TIR radiation. This result shows the possibility that the thermal infrared data can be used reasonably well to calibrate brightness temperature if high resolution real time data are available.

A model has been constructed for predicting temperature over the study area, the details of which are given in Chapter 6. The predicted temperature over 85% of the study area has standard deviation error less than 1.0, irrespective of the month or season and thus justifies the model construction and selection of average weights for different variable for predicting mean temperature. The standard deviation is less in monsoon and more in winter season while it is in between for both premonsoon and postmonsoon seasons. Seasonal wind pattern changes from the southerly to northerly in winter season. Dry and cool northerly wind blows over the study area from the land-locked regions of upper latitudes beyond Himalayas with the southward swinging Intertropical Convergence Zone (ITCZ). In addition, extra-tropical disturbances (ETD) pass through the Himalayan region and while passing through the north of the study area, induced low pressure system form over the study area. As a result parts of the area receives small amount of winter rainfall dropping the temperature dramatically from faster latent cooling induced from dry ambience and clear sky, thereby leading to severe cold wave conditions. On the other hand, monsoon being a synoptic scale system, the entire study area receives uniform impact of monsoon on an average basis. Therefore the study suggests that general wind pattern is to be considered as another independent variable which has not been considered in the present study. A direct one-to-one comparison of predicted values and IMD observed values reveals that the difference in mean monthly temperature is within 1 degree C for almost all stations irrespective of the months in premonsoon season. The same comparison can be carried out for other seasons as the observed data for other seasons can be made available. This is very much required also for justifying the acceptability of the model for the region and for all the seasons. Similar model has been constructed for rainfall modeling using different independent variables. In rainfall modelling, standard deviation errors are less than 10 over a large part of the area, irrespective of the month or season, justifying the applicability of the procedure used for predicting total rainfall. However, it is striking to note that standard deviation errors are more in the monsoon region compared to the other seasons. Possibly it is because of interannual variability of monsoonal rainfall over the study area. Again as the amount of rainfall is very high in the monsoon season compared to any other seasons, the variability is also high leading to higher standard deviation errors. Similar comparison like temperature, i.e., comparison of predicted rainfall with observed values for premonsoon season reveals that the predicted values are close to the observed values except a few stations where on some occasions prediction is poor. Thus climatological modeling shows an effective tool for generating high-resolution climate data sets, although the present methodology requires further improvement for better accuracy.

8.2 Recommendations for future studies

Indeed, the present work makes an initial attempt for climatological modeling, particularly temperature and rainfall modeling over a tropical region, but it also evokes lot of scopes for further studies. Some of the ideas are presented here. Of course there is every scope for further improvements in land use/ land cover characterization as well as using DEM data built from higher resolution satellite data. Emphasis might be placed in constructing the model and weight selection for different independent variables. Variables such as 'distance from the coast line', 'distance from large inland water bodies', prevailing global and local wind patterns etc might be considered for further improvements.

Appendixes

Appendix 1. Identification of different land use feature in different points

Sl. No.	Spatial location		Observe feature
	Longitude (°E)	Latitude(°N)	
1	87°09′51.58″	22°53′56.97″	Agricultural land
2	87°41′01.44″	22°39′38.65″	Agricultural land
3	87°22′19.04″	22°27′54.25″	Agricultural land
4	87°04′25.85″	23°00′30.85″	Agricultural land
5	87°27′00.34″	22°28′33.42″	Agricultural land
6	86°55′03.84″	23°07′05.77″	Agricultural land
7	87°15′25.21″	23°01′12.30″	Forest
8	87°20′54.43″	22°49′07.55″	Forest
9	87°22′06.67″	22°39′17.57″	Forest
10	87°16′07.68″	22°43′50.06″	Forest
11	87°16′15.31″	22°45′11.83″	Forest
12	87°07′45.12″	22°47′46.77″	Forest
13	87°12′54.99″	22°44′33.53″	Forest
14	86°04′32.90″	23°11′10.40″	Agriculture fallow
15	86°04′47.50″	23°11′13.50″	Cultivated land
16	86°04′14.90″	23°11′12.20″	Agriculture fallow
17	86°04′54.10″	23°11′12.60″	Water body
18	86°04′57.00″	23°11′19.70″	Agriculture fallow
19	87°59′25.23″	22°53′37.14″	Forest
20	87°17′23.63″	22°29′28.94″	Forest
21	87°16′26.36″	22°29′13.47″	Forest
22	87°12′04.18″	22°31′58.05″	Forest
23	87°10′07.35″	22°38′12.69″	Forest
24	86°04′40.01″	23°11′30.20″	Water body
25	86°04′41.60″	23°11′09.40″	Water body
26	86°04′32.90″	23°11′09.20″	Agriculture fallow
27	87°17′05.76″	22°43′09.68″	Forest
28	87°16′43.37″	22°43′24.33″	Forest
29	86°28′57.86″	22°35′43.92″	Agriculture fallow
30	86°48′16.74″	22°31′38.24″	Agriculture fallow
31	87°00′32.28″	22°52′14.10″	Forest
32	86°04′55.00″	23°11′20.00″	vegetation
33	86°04′38.10″	23°11′13.00″	Agriculture fallow
34	86°07′39.20″	23°12′47.50″	Agriculture fallow
35	86°07′20.40″	23°12′59.90″	Water body
36	86°07′07.41″	23°13′02.61″	Agriculture fallow
37	86°06′53.01″	23°12′46.21″	forest
38	86°06′44.61″	23°12′33.51″	forest
39	86°06′36.01″	23°12′32.91″	forest
40	86°06′24.41″	23°12′28.61″	Water body
41	85°04′32.16″	22°50′30.14″	Agricultural land
42	86°04′22.91″	22°57′31.31″	Agricultural land
43	85°37′09.96″	22°40′42.25″	Agricultural land
44	87°06′11.28″	22°56′29.82″	Forest
45	86°59′33.37″	23°52′15.32″	Forest
46	87°01′26.88″	22°41′52.45″	Agriculture fallow
47	87°03′12.24″	22°29′17.85″	Agricultural land
48	87°01′26.88″	22°22′16.68″	Agricultural land
49	87°22′13.56″	22°30′28.05″	Agricultural land
50	87°32′45.67″	22°21′24.03″	Agricultural land

51	87°54'07.46"	22°18'46.09"	Agriculture fallow
52	87°28'22.29"	22°05'53.95"	Agriculture fallow
53	87°31'52.99"	21°54'29.54"	Agriculture fallow
54	87°49'49.76"	23°40'59.23"	Open Shrub
55	86°46'13.83"	21°57'07.48"	Agricultural land
56	86°47'59.18"	21°32'15.84"	Agricultural land
57	86°30'25.66"	21°05'03.80"	Agricultural land
58	86°10'31.66"	20°57'45.08"	Agricultural land
59	85°35'59.73"	20°39'54.60"	Agricultural land
60	85°42'23.12"	22°30'44.54"	Open Shrub
61	86°25'09.60"	20°31'25.69"	Agricultural land
62	85°50'37.66"	20°11'32.37"	Agricultural land
63	88°08'29.21"	21°15'29.51"	Agriculture fallow
64	88°05'26.80"	21°45'09.00"	Agriculture fallow
65	88°04'45.06"	21°39'57.50"	Agriculture fallow
66	88°11'38.10"	21°52'27.81"	Agriculture fallow
67	88°1414.50"	21°46'03.40"	Agricultural land
68	88°15'00.20"	21°41'44.30"	Agricultural land
69	88°15'20.21"	21°37'52.30"	Agricultural land
70	88°16'02.08"	21°33'54.20"	Agricultural land
71	88°18'58.00"	21°50'22.31"	Agricultural land
72	88°18'37.20"	21°50'38.31"	Agricultural land
73	85°05'45.74"	23°53'01.74"	Open Shrub
74	88°23'56.89"	24°36'10.76"	Open Shrub
75	87°48'49.55"	23°06'21.99"	Open Shrub
76	88°46'01.50"	23°38'58.81"	Open Shrub
77	85°59'56.76"	22°28'14.02"	Open Shrub
78	85°01'44.62"	24°13'05.94"	Open Shrub
79	85°36'21.86"	23°39'59.02"	Open Shrub
80	87°01'09.58"	24°02'33.73"	Open Shrub
81	88°53'33.08"	23°06'21.99"	Fallow land
82	87°45'18.81"	22°12'10.66"	Fallow land
83	88°03'52.69"	24°05'04.26"	Fallow land
84	88°20'26.15"	24°56'14.96"	Fallow land
85	88°59'04.23"	24°34'10.34"	Fallow land
86	88°12'24.48"	24°54'14.54"	Fallow land
87	88°47'31.82"	22°37'46.01"	Fallow land
88	88°34'09.82"	23°40'47.92"	Fallow land
89	87°52'54.66"	23°46'54.61"	Fallow land
90	87°37'07.38"	22°34'35.45"	Fallow land
91	88°21'25.88"	22°36'07.12"	Settlement
92	87°19'48.42"	22°18'48.16"	Settlement
93	87°19'17.86"	23°04'38.34"	Settlement
94	87°53'25.22"	23°13'48.37"	Settlement
95	88°14'18.07"	24°14'24.72"	Settlement
96	85°18'05.18"	23°20'56.18"	Settlement
97	86°12'34.83"	22°46'18.27"	Settlement
98	87°13'41.73"	24°59'44.33"	Settlement
99	88°08'11.38"	24°50'34.30"	Settlement
100	86°22'15.42"	23°19'24.51"	Settlement

Appendix 2. Classification accuracy assessment report for land use/ land cover

Appendix 2.1
Pixel Size – 30m x 30m

Image File: e:/phd/work/phd_data set_21-25-85-89/landuse_30m.img
User Name: Administrator
Date: Fri Oct 23 07:25:22 2010

ERROR MATRIX   Reference Data

Classified Data	Water	Marshy	Forest	Shrub	Agriculture	Urban & built-up
Water	5	0	0	0	0	0
Marshy	0	4	0	0	0	0
Forest	0	0	6	0	0	0
Shrub	0	0	1	5	1	0
Agriculture	0	0	1	2	22	0
Urban & built-up	0	0	0	0	0	3
Column Total	**5**	**4**	**8**	**7**	**23**	**3**

ACCURACY TOTALS

Class Name	Reference Totals	Classified Totals	Number Correct	Producers Accuracy	Users Accuracy
Water	5	5	5	100.00%	100.00%
Marshy	4	4	4	100.00%	100.00%
Forest	8	6	6	75.00%	100.00%
Shrub	7	7	5	71.43%	71.43%
Agriculture	23	25	22	95.65%	88.00%
Urban & built-up	3	3	3	100.00%	100.00%
Totals	**50**	**50**	**45**		

Overall Classification Accuracy =   90.00%

Appendix 2.2

Pixel Size - 1km x 1km

Image File: e:/phd/work/phd_data set_21-25-85-89/landuse_1km.img
User Name: Administrator
Date: Fri Oct 25 07:28:22 2010

ERROR MATRIX   Reference Data

Classified Data	Water	Marshy	Forest	Shrub	Agriculture	Urban & built-up
Water	5	0	0	0	0	0
Marshy	0	4	0	0	0	0
Forest	0	0	6	1	0	0
Shrub	0	0	1	4	2	0
Agriculture	0	0	1	2	21	0
Urban & built-up	0	0	0	0	0	3
Column Total	**5**	**4**	**8**	**7**	**23**	**3**

ACCURACY TOTALS

Class Name	Reference Totals	Classified Totals	Number Correct	Producers Accuracy	Users Accuracy
Water	5	5	5	100.00%	100.00%
Marshy	4	4	4	100.00%	100.00%
Forest	8	7	6	75.00%	85.71%
Shrub	7	7	4	57.14%	57.14%
Agriculture	23	24	21	91.30%	87.50%
Urban & built-up	3	3	3	100.00%	100.00%
Totals	**50**	**50**	**43**		

Overall Classification Accuracy =   86.00%

Appendix 3. Identification of different soil texture in different sample location

Sample id	Longitude(°E)	Latitude (°N)	Place Name	% sand	% silt	% clay	Soil Texture
1	87°11′42.36″	22°26′15.00″	Bager Pukur_Kotoali Thana	50	30	20	Sandy Clay Loam
2	86°56′53.52″	22°32′28.20″	Andhariya_Binpur′1	50	30	20	Sandy Clay Loam
3	87°03′34.56″	22°29′12.84″	Sundrabari_Lalgarh	40	50	10	Silt Loam
4	86°50′40.92″	22°36′05.76″	Silda GP_Silda	40	40	20	Sandy Loam
5	86°49′0.48″	22°31′17.76″	Dharsa_Jamboni	30	40	30	Sandy Loam
6	86°53′54.24″	22°30′14.04″	Kapgeria	40	40	20	Sandy Loam
7	87°01′35.40″	22°22′06.24″	Salbani, Jhargram	40	50	10	Silty Loam
8	87°20′9.96″	22°21′05.04″	Near Chourangi, KGP	10	50	40	Silty Clay
9	87°54′36.00″	22°18′16.20″	Maniktala	40	40	20	Sandy Loam
10	88°00′52.92″	22°15′25.20″	Pururpara, Bagnan	0	20	80	Clay
11	87°44′45.96″	22°23′53.68″	Panskura, Mechogram	20	50	30	Silty Clay
12	87°33′04.68″	22°21′31.32″	Balichack	20	30	50	Silty Clay
13	86°04′37.56″	22°57′20.88″	Ckandil	30	50	20	Silty Clay Loam
14	86°13′22.44″	22°50′24.72″	Dimna_Jamsedpur	40	40	20	Sandy Loam
15	85°12′08.00″	22°22′27.12″	Monoharpur	80	20	0	Sandy
16	85°48′27.72″	22°31′56.34″	Chibasa	20	20	60	Clay Loam
17	85°45′52.92″	22°24′31.68″	Jiglami	10	20	70	Clay Loam
18	85°37′31.08″	22°40′33.60″	Chakradharpur	80	20	0	Sandy
19	85°20′04.92″	22°51′29.52″	Bandgoon	80	10	10	Sandy
20	85°48′49.68″	22°44′32.28″	Kharswan	20	20	60	Clay Loam
21	85°04′44.04″	22°50′20.04″	Kandra	50	30	20	Sandy Clay Loam
22	86°29′32.64″	22°35′13.92″	Ghatshila	50	30	20	Sandy Loam
23	85°50′58.56″	20°11′31.56″	Dhouli	40	50	10	Silty Loam
24	85°48′05.04″	20°15′47.16″	Barmunda	40	50	10	Silty Loam
25	85°36′24.84″	20°39′22.10″	Dhankanal	50	30	20	Sandy Clay Loam
26	86°25′32.12″	20°30′51.84″	Kendrapara	10	20	70	Clay Loam
27	86°15′21.60″	20°29′13.56″	Asureshar	10	20	70	Clay Loam
28	86°06′38.88″	20°28′33.24″	Salipur	20	40	40	Silty Loam
29	86°44′51.72″	21°07′54.48″	Basudevpur	80	20	0	Sandy
30	86°10′51.96″	20°57′23.76″	Korai	20	20	60	Clay Loam
31	86°30′47.88″	21°04′46.56″	Bhadrak	50	30	20	Sandy Clay Loam
32	86°46′35.40″	21°56′58.56″	Baripada	40	50	10	Silty Loam
33	86°54′26.28″	21°30′06.84″	Balasore	20	30	50	Silty Clay
34	86°48′25.20″	21°31′46.56″	Chandipur	10	50	40	Silty Clay
35	87°01′52.32″	22°41′42.72″	Sarenga_Brahmundiha	10	20	70	Clay Loam
36	87°34′10.92″	22°13′23.88″	Temathani	10	20	70	Clay Loam
37	87°23′46.68″	22°12′37.08″	Makrampur	0	20	80	Clay
38	87°20′35.88″	22°04′20.28″	Belda	10	20	70	Clay Loam
39	87°32′12.12″	21°54′03.60″	Egra	10	40	50	Clay Loam
40	87°31′51.24″	22°01′07.32″	Pataspur	20	30	50	Silty Clay
41	87°22′55.20″	22°30′19.80″	Goilkera	20	30	50	Clay Loam
42	87°28′54.48″	22°05′34.80″	Dehati	10	20	70	Clay Loam
43	88°05′26.88″	21°45′09.36″	Harinbari, Sagar	10	70	20	Silty Loam
44	88°4′45.84″	21°39′56.88″	Light House, Sagar	10	60	30	Silty Loam
45	88°11′39.84″	21°52′27.48″	Kakdwip	10	75	15	Silty Loam
46	88°14′14.24″	21°46′03.36″	Namkhana	5	70	25	Silty Clay Loam
47	88°15′00.36″	21°41′44.16″	S Chandranagar, Namkhana	10	65	25	Silty Clay Loam
48	88°15′20.52″	21°37′51.60″	10MilePatibenia, Namkhana	5	75	20	Silty Loam
49	88°18′31.32″	21°50′58.92″	Durbachatti, Patharpratim	5	75	20	Silty Loam

Appendix 4. Soil texture Classification according to their different characteristics

Soil Type	Map Symbol	Soil Type	Taxonomic Name
Loamy Sand	W6	Coarse loamy	*Umbric Dystrochrepts*
	W8	Coarse loamy	*Typic Haplaquent*
	W9	Coarse loamy	*Aquic Udifluvents*
	W10,W18, W22 ,W25	Coarse loamy	*Aquic Ustifluvents*
	W30	Coarse loamy	*Typic Ustorthents*
	W60	Coarse loamy	*Typic Fluvaquents*
	W15,W61,W63 ,W64	Coarse loamy	*Typic Ustifluvents*
	W67	Coarse loamy	*Typic Haplustalfs*
	1	Younger Alluvial soil	*Udifluvents*
Sandy Loam	W2	Coarse loamy	*Typic Dystrochrepts*
	W5	Coarse loamy	*Typic Udorthents*
	W17	Coarse loamy	*Typic Ustorthents*
	W107	Coarse loamy	*Typic Ustifluvents*
Silty Loam	W14	Fine	*Typic Ustochrepts*
	W16	Fine silty	*Typic Ustifluvents*
	W28	Fine silty	*Typic Fluvaquents*
Loamy	W92	Loamy	*Ustorthents*
	W96	Loamy	*Lithic Haplustalts*
	W99	Loamy	*Lithic Haplustalts*
	W93, W95, W105, W109	Loamy	*Lithic Ustorthents*
	W110	Loamy	*Lithic Ustochrepts*
	W113,W114, W115	Loamy	*Lithic Haplustalts*
	10	Shallow black soil	*Ustochrepts*
	22	Red Loamy	*Haplustalfs,Paleustalfs, Rhodustalfs*
Sandy Clay Loam	W3	Fine loamy	*Umbric Dystrochrepts*
	W7	Fine loamy	*Fluventic Eutrochrepts*
	W11,W12, W13	Fine loamy	*Typic Haplaquents*
	W24	Fine loamy	*Fluventic Ustochrepts*
	W31 ,W32, W33, W66, W101, W106	Fine loamy	*Typic Ustochrepts*
	W19,W27, W34,W35,W53	Fine loamy	*Fluventic Ustochrepts*
	W62	Fine loamy	*Aquic Ustifluvents*
	W65	Fine loamy	*Typic Ustifluvents*
	W68	Fine loamy	*Ultic Paleustalfs*
	W103	Fine loamy	*Rhodic Paleustalfs*
	W104	Fine	*Typic Paleustalfs*
	W79,W80, W91,W94, W97,W98, W100,W102, W108,W111, W112	Fine loamy	*Typic Haplaquents*
	23	Red Earths	*Haplustalfs,Paleustalfs, Rhodustalfs*
	24	Mixed Red and Black soil	*Rhodustalfs, Pallusterts*
	25	Brown, Red &Black soil	*Palehumults*
	26	Red and Yellow soil	*Ochraqults, Rhodustalfs, Haplustalfs*
	27,28	Lateritic soil	*Plinthaquenlts,*

			Plinthustults, Plinthudults
Silty Clay Loam	W1	Loamy Skeletal	Lithic Udorthents
	W4	Loamy Skeletal	Typic Haplumbrepts
	W20	Fine loamy	Typic Ustochrepts
	W26,W29, W50, W51,W82	Fine loamy	Aeric Haplaquepts
	W40 ,W43	Fine	Vertic Ochraqualfs
	W41,W44, W45	Fine	Vertic Haplaquepts
	W58, W59	Fine	Aeric Haplaquepts
	W69	Fine loamy	AericOchraqualfs
	W70	Fine	Aeric Ochraqualfs
	W90	Fine loamy	Typic Haplaquepts
	14	Saline and Saline Alkali soils	Natrargids, Salargids
	21	Red Sandy soil	Haplustalfs, Paleustalfs, Rhodustalfs
Clay Loam	W19,W23, W54	Fine loamy	Typic Ustochrepts
	W71	Fine loamy	Typic Haplustalfs
	W72	Fine	Typic Haplustalfs
Sandy clay	W21,W75,W76	Fine	Aeric Haplaquepts
	2	Coastal Alluvial soil	Haplaquents
	19	Older Alluvial soil	Paleustalfs, Haplaquents
Silty Clay	W36	Fine	Vertic Ochraqualfs
	W38,W47	Very Fine	Aeric Haplaquepts
	W46,W49, W52, W56, W77	Fine	Typic Haplaquepts
	W78	Fine	Vertic Haplaquepts
	W37,W42, W48, W55, W81, W83, W84,W85, W86, W87, W88, W89	Fine	Aeric Haplaquepts

Source: West Bengal Soil Type map, National Bureau of soil survey & Land Use Planning and India Soil Region map, National Atlas & thematic Mapping Organization Govt. of India.

Note:

Map symbol W1-115: WB Soil Type, National Bureau of soil survey & Land Use Planning
Map symbol 0 -30: India Soil Region, National Atlas & thematic Mapping Organization

Appendix 5. Interpolated mean temperature surfaces from IDWA, Splining, and Kriging for June

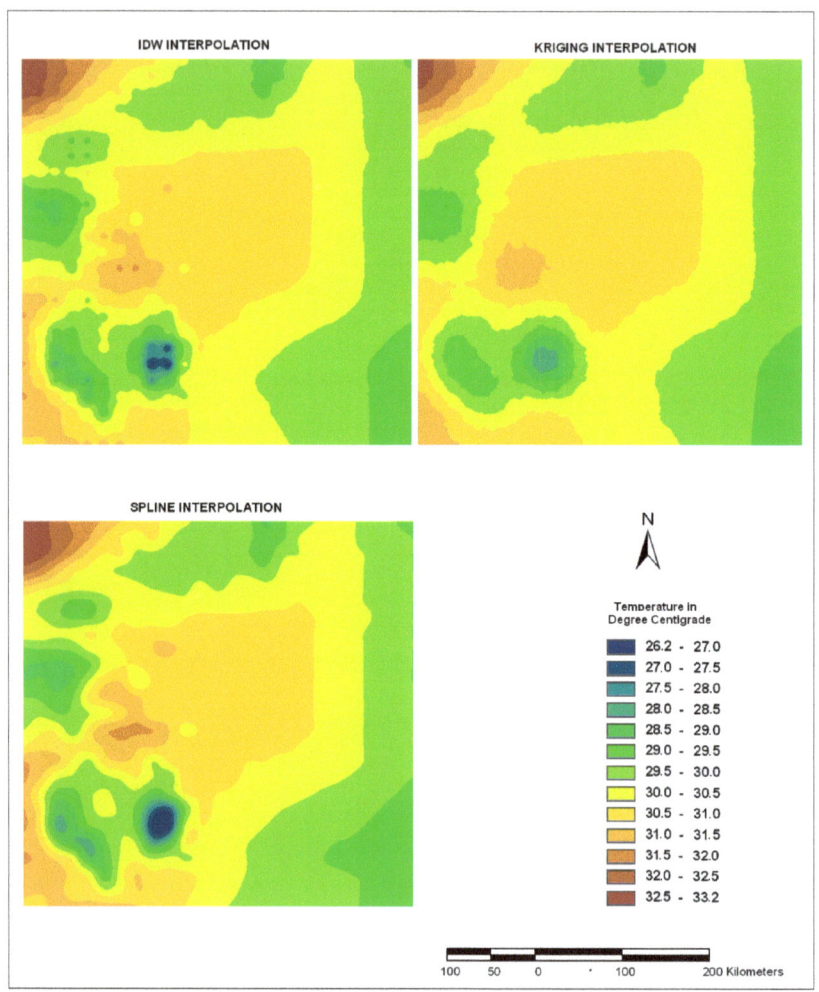

Appendix 6. Interpolated total monthly rainfall surfaces from IDWA, Splining, and Kriging for June

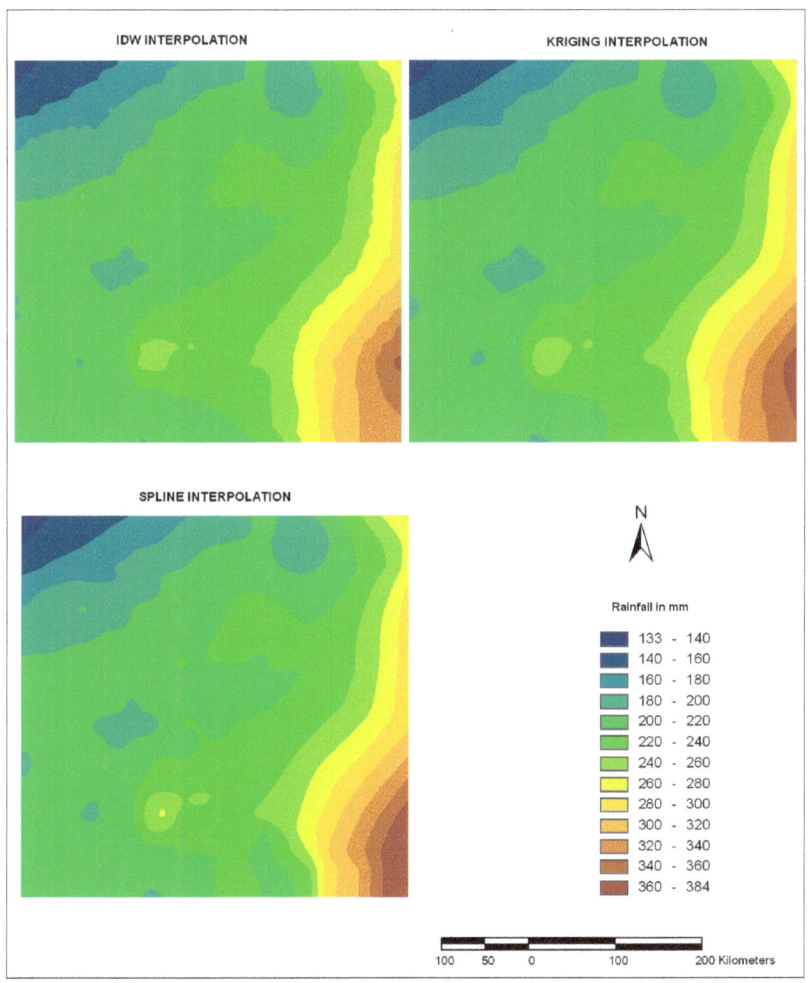

Appendix 7. Comparison of CRU data set with IMD observed values for premonsoon season monthly mean temperature

Station-1	Digha			Station-11	Diamondharbour		
Monthly Temperature	March	April	May	Monthly Temperature	March	April	May
IMD data set	26.1	28.4	29.2	IMD data set	27.2	29.5	30.0
CRU	27.6	30.3	31.4	CRU	27.4	29.6	30.8
Station-2	Balasore			Station-12	Contai		
IMD data set	27.3	29.8	30.5	IMD data set	26.9	29.3	29.5
CRU	27.8	30.3	31.4	CRU	27.6	30.3	31.4
Station-3	Midnapur			Station-13	Dumdum		
IMD data set	27.8	31.2	31.4	IMD data set	27.1	30.1	30.4
CRU	27.4	30.8	32.3	CRU	27.6	30.1	31.1
Station-4	Ranchi			Station-14	Durgapur		
IMD data set	24.2	28.1	30.1	IMD data set	23.4	30.8	31.2
CRU	24.9	29.2	30.9	CRU	24.7	30.7	32.3
Station-5	Bankura			Station-15	Uluberia		
IMD data set	29.0	29.7	32.8	IMD data set	26.9	29.8	30.2
CRU	27.1	30.9	32.4	CRU	27.9	30.6	31.5
Station-6	Jamshedpur			Station-16	Burdwan		
IMD data set	27.3	30.9	32.7	IMD data set	27.5	31.2	32.0
CRU	26.1	29.9	31.9	CRU	27.9	31.1	32.0
Station-7	Chaibasa			Station-17	Santiniketan		
IMD data set	26.4	30.4	31.6	IMD data set	26.3	30.3	30.1
CRU	25.6	29.7	31.9	CRU	27.2	31.0	32.2
Station-8	Krishnanagar			Station-18	Baripada		
IMD data set	25.7	29.7	30.0	IMD data set	27.5	31.1	31.1
CRU	27.7	30.7	31.5	CRU	27.8	30.5	31.8
Station-9	Purilia			Station-19	Haldia		
IMD data set	25.6	29.7	30.9	IMD data set	27.0	29.3	29.7
CRU	25.9	30.0	32.3	CRU	27.4	29.6	30.8
Station-10	Alipur			Station-20	Hazaribag		
IMD data set	27.6	30.3	30.8	IMD data set	23.3	27.9	29.9
CRU	27.9	30.6	31.5	CRU	23.6	28.2	30.3

Note:

IMD data set: Observed temperature data set of India Meteorological Department
CRU: Climate Research Unit data set

Appendix 8. Comparison of CRU data set with IMD observed values for premonsoon season monthly total rainfall

Station-1	Digha			Station-11	Diamondharbour		
Monthly Rainfall	March	April	May	Monthly Rainfall	March	April	May
IMD data set	31.8	30.5	97.1	IMD data set	27.3	49.9	108.7
CRU	26.2	68.4	147.2	CRU	35.1	65.5	107.2
Station-2	Balasore			Station-12	Contai		
IMD data set	51.7	80.9	127.8	IMD data set	34.4	41.7	74.3
CRU	38.0	68.1	111.3	CRU	26.2	68.4	147.2
Station-3	Midnapur			Station-13	Dumdum		
IMD data set	52.4	68.4	139.4	IMD data set	22.9	45.8	113.3
CRU	34.2	55.6	100.6	CRU	33.2	65.0	125.4
Station-4	Ranchi			Station-14	Asansol		
IMD data set	21.9	21.2	60.8	IMD data set	27.3	22.0	78.0
CRU	18.8	15.6	39.6	CRU	25.4	36.0	74.5
Station-5	Bankura			Station-15	Uluberia		
IMD data set	35.9	38.8	88.4	IMD data set	19.8	42.9	85.0
CRU	29.5	45.3	85.1	CRU	37.1	67.0	106.3
Station-6	Jamshedpur			Station-16	Burdwan		
IMD data set	28.9	28.1	67.1	IMD data set	39.4	42.7	85.4
CRU	23.8	30.3	72.3	CRU	35.8	61.6	108.9
Station-7	Chaibasa			Station-17	Santiniketan		
IMD data set	20.5	31.7	69.8	IMD data set	38.0	44.0	90.5
CRU	21.5	23.0	58.6	CRU	32.8	46.2	93.8
Station-8	Krishnanagar			Station-18	Baripada		
IMD data set	19.2	44.6	62.6	IMD data set	33.5	38.7	98.9
CRU	34.5	72.9	126.1	CRU	35.7	53.3	104.7
Station-9	Purilia			Station-19	Haldia		
IMD data set	29.1	35.9	70.4	IMD data set	27.3	49.9	95.5
CRU	20.0	25.2	62.2	CRU	35.1	65.5	107.2
Station-10	Alipur			Station-20	Hazaribag		
IMD data set	22.9	49.3	103.7	IMD data set	17.7	16.1	53.4
CRU	37.1	67.0	106.3	CRU	14.6	11.9	33.3

Note:
  IMD data set: Observed rainfall data set of India Meteorological Department
  CRU: Climate Research Unit data set

Appendix 9. Modeled temperature against selected parameters for all months (1×10^{-1})

Land use	Weight in %	January			February			March			April		
		30 Years	20 Years	10 Years	30 Years	20 Years	10 Years	30 Years	20 Years	10 Years	30 Years	20 Years	10 Years
Forest	33	65.86	68.18	68.41	76.29	78.96	79.26	94.67	96.83	97.16	114.01	113.88	113.77
Shrub	34	70.43	70.64	70.79	80.65	81.35	81.79	99.13	99.44	99.73	110.83	112.26	112.12
Crop	36	71.91	74.07	74.22	82.86	85.48	85.78	102.36	104.83	105.09	118.10	155.00	119.91
Marshy	38	78.62	79.85	79.99	90.07	92.62	92.81	110.03	111.76	111.49	123.22	124.22	123.86
Urban	40	82.56	82.90	83.18	94.89	95.87	96.08	115.47	116.24	116.38	129.00	129.57	129.57
Soil Texture													
1	20	38.30	38.46	38.57	44.14	44.70	44.88	54.93	55.38	55.38	63.81	64.07	63.91
2	19	36.67	38.45	38.56	42.57	44.61	44.74	52.96	54.80	54.99	62.28	63.61	63.51
3	18	36.19	36.19	36.26	41.17	41.68	41.81	50.17	50.49	50.36	58.36	57.55	57.43
4	17	35.50	35.45	35.56	40.86	40.65	40.89	49.90	49.66	49.87	57.32	56.78	56.78
5	16	32.94	33.94	34.03	37.86	39.02	39.21	46.74	47.63	47.83	54.04	55.00	54.64
6	15	31.53	31.80	31.83	36.25	36.83	36.91	44.34	44.74	44.69	49.92	50.23	50.10
7	14	30.01	30.54	30.62	34.31	35.25	35.33	41.72	42.36	42.40	47.17	47.26	47.18
8	13	26.68	26.88	27.00	30.50	30.97	31.09	36.66	36.49	36.46	43.42	43.07	42.90
9	12	25.00	24.68	24.77	28.98	29.04	29.12	36.65	37.10	37.19	40.45	40.99	41.06
Elevation(m)													
1100	30	59.86	61.40	62.80	69.76	72.57	73.33	86.52	91.95	90.71	99.10	110.14	110.52
1000	31	61.36	62.90	64.30	71.26	74.07	74.83	88.02	93.45	92.21	100.60	111.64	112.02
900	32	62.86	64.40	65.80	72.76	75.57	76.33	89.52	94.95	93.71	102.1	113.1	113.52
800	33	64.97	66.60	67.01	74.77	77.58	78.23	92.17	96.86	95.50	104.93	114.23	114.49
700	34	67.25	68.81	69.01	77.92	79.00	80.18	96.61	98.22	98.67	112.74	116.03	117.05
600	35	68.16	70.74	71.84	79.14	81.87	82.07	98.92	101.45	101.75	116.93	118.69	118.39
500	36	68.87	72.05	73.26	80.15	84.86	84.82	99.99	104.18	104.47	118.45	121.38	121.05
400	37	74.87	74.64	74.80	86.25	86.29	86.66	107.15	106.30	106.67	125.88	123.78	123.57
300	38	79.05	79.19	79.46	90.48	91.20	91.74	111.51	111.43	111.92	129.13	128.05	127.94
200	39	79.88	80.55	80.84	92.10	92.72	93.21	113.34	113.70	114.13	131.30	131.02	130.90
100	40	82.62	82.87	83.06	94.82	96.10	96.41	116.09	116.93	116.90	131.11	131.69	131.49

Land use	Weight in %	May			June			July			August		
		30 Years	20 Years	10 Years	30 Years	20 Years	10 Years	30 Years	20 Years	10 Years	30 Years	20 Years	10 Years
Forest	33	116.84	118.86	120.18	112.46	114.00	115.17	102.41	103.37	103.97	100.55	101.79	102.18
Shrub	34	118.70	120.78	122.13	115.10	114.66	115.35	106.96	106.01	106.60	105.96	105.10	105.25
Crop	36	123.29	125.49	126.85	120.42	121.46	122.12	112.75	112.77	113.17	111.67	111.75	111.82
Marshy	38	125.19	126.61	127.47	123.09	124.41	124.23	119.28	120.15	120.65	118.82	119.91	119.78
Urban	40	132.02	132.95	134.36	128.58	129.22	129.43	122.59	122.88	123.02	121.47	121.89	121.68
Soil Texture													
1	20	67.07	67.53	68.27	64.78	65.18	65.49	60.90	60.84	61.08	60.31	60.14	60.12
2	19	66.21	66.83	67.49	63.85	64.76	65.41	58.24	58.64	58.95	57.41	57.81	57.94
3	18	59.45	59.18	59.95	57.29	57.45	57.45	55.47	55.53	55.79	55.34	55.34	55.28
4	17	57.57	58.02	58.6	57.21	56.91	57.37	52.78	52.46	52.73	52.24	51.90	52.03
5	16	56.67	57.47	58.13	54.54	55.11	55.58	49.99	50.32	50.62	49.45	49.82	49.94
6	15	50.89	51.45	51.99	49.70	50.07	50.09	47.62	47.71	47.86	47.33	47.61	47.54
7	14	48.53	48.69	48.82	47.11	47.27	47.32	44.56	44.68	44.88	44.20	44.37	44.41
8	13	45.21	45.21	45.65	44.95	44.48	44.71	42.51	41.46	41.60	42.10	40.93	40.87
9	12	41.38	41.92	42.37	40.43	40.79	40.83	38.71	38.93	39.03	38.40	38.67	38.63
Elevation(m)													
1100	30	103.29	113.95	115.48	98.71	109.00	110.14	90.71	99.86	101.00	89.57	99.10	99.48
1000	31	104.79	115.45	116.98	100.21	110.50	111.64	92.21	101.36	102.50	91.07	100.60	100.98
900	32	106.29	116.95	118.48	101.71	112.00	113.14	93.71	102.86	104.00	92.57	102.10	102.48
800	33	109.44	118.61	120.29	104.80	113.84	114.62	96.68	105.46	106.24	95.78	104.43	104.95
700	34	119.59	124.49	125.89	114.41	118.67	120.01	103.31	106.82	107.52	101.92	105.38	105.63
600	35	124.36	126.99	128.30	119.20	121.07	122.48	107.21	108.62	109.08	105.54	107.05	107.30
500	36	126.88	129.69	130.95	123.39	124.15	125.46	112.60	111.70	112.20	110.79	110.15	110.39
400	37	133.38	131.80	133.17	128.08	125.97	127.39	115.69	113.54	114.22	114.12	112.01	112.33
300	38	135.34	134.15	135.86	130.36	128.95	130.08	119.66	117.88	118.63	118.27	116.60	116.87
200	39	136.43	136.78	138.45	132.39	132.24	133.22	122.68	121.90	122.48	121.52	120.54	120.71
100	40	137.80	138.84	140.03	135.19	135.38	135.61	126.17	126.07	126.51	125.36	125.34	125.23
Weight		September			October			November			December		

Land use	in %	30 Years	20 Years	10 Years	30 Years	20 Years	10 Years	30 Years	20 Years	10 Years	30 Years	20 Years	10 Years
Forest	33	100.57	102.03	101.89	94.22	96.35	97.83	79.71	82.05	83.64	66.65	70.08	70.71
Shrub	34	105.73	105.14	105.04	99.83	99.63	100.74	85.22	85.41	86.64	71.42	72.73	73.24
Crop	36	111.12	111.56	111.43	104.43	105.88	107.00	88.58	90.58	91.86	73.55	76.69	77.15
Marshy	38	119.09	119.64	119.68	113.87	114.82	115.60	98.94	100.46	101.68	81.92	84.61	84.88
Urban	40	121.54	121.61	121.25	116.05	116.04	116.60	100.47	100.60	101.87	83.77	85.65	85.93
Soil Texture													
1	20	60.09	60.12	59.98	56.63	56.93	57.44	48.10	48.53	49.21	39.66	40.62	40.86
2	19	57.01	57.84	57.75	53.14	54.55	55.45	46.68	46.38	47.29	38.20	39.58	39.96
3	18	55.28	55.21	55.28	52.79	53.11	53.43	45.70	45.40	46.85	38.11	38.87	39.06
4	17	52.14	52.01	51.96	49.35	49.33	49.78	42.15	42.10	42.68	35.56	36.04	36.20
5	16	49.30	49.83	49.80	46.35	47.13	47.79	39.44	40.36	41.06	33.13	34.60	34.89
6	15	47.40	47.49	47.43	45.18	45.38	45.71	39.01	39.49	39.85	32.33	33.20	33.27
7	14	44.43	44.49	44.41	42.43	42.53	42.92	36.85	37.26	37.73	30.89	31.77	31.89
8	13	41.75	40.70	40.67	39.03	38.59	38.91	32.66	32.30	32.74	27.10	27.68	27.79
9	12	38.47	38.58	38.54	36.89	36.96	37.14	32.23	32.24	32.72	26.52	26.61	26.84
Elevation(m)													
1100	30	89.57	99.10	99.48	84.24	94.14	95.29	71.29	81.57	83.10	59.86	70.52	71.29
1000	31	91.07	100.60	100.98	85.74	95.64	96.79	72.79	83.07	84.60	61.36	72.02	72.79
900	32	92.57	102.10	102.48	87.24	97.14	98.29	74.29	84.57	86.10	62.86	73.52	74.29
800	33	95.65	104.56	104.56	90.11	99.15	100.31	76.70	85.00	86.42	64.97	74.40	75.04
700	34	101.48	105.53	105.43	94.81	99.60	101.29	80.46	86.00	86.85	67.59	75.04	75.84
600	35	104.94	107.05	106.80	97.86	100.63	102.60	82.32	87.20	87.08	68.77	75.81	76.57
500	36	109.54	110.10	109.86	101.20	103.74	105.72	84.57	87.91	89.76	69.93	76.10	76.84
400	37	113.70	112.02	111.86	106.35	105.45	107.29	89.92	89.60	91.49	75.33	76.64	77.37
300	38	118.06	116.81	116.70	111.22	110.35	111.64	94.30	94.44	96.05	79.41	80.78	81.43
200	39	121.00	120.59	120.45	114.17	114.35	115.42	96.96	97.37	98.57	80.97	82.67	83.14
100	40	125.38	125.10	124.98	119.39	119.51	120.30	102.97	103.57	104.69	85.47	87.20	87.53

Appendix 10. Modeled rainfall against selected parameters for different months (1×10^{-1})

Land use	Weight in %	May 30 Years	May 20 Years	May 10 Years	June 30 Years	June 20 Years	June 10 Years	July 30 Years	July 20 Years	July 10 Years	August 30 Years	August 20 Years	August 10 Years
Forest	5	123	120	120	174	166	166	169	162	162	177	175	175
Shrub	4	91	116	116	115	111	111	131	119	119	127	131	131
Crop	3	70	84	84	91	92	92	104	99	99	97	101	101
Marshy	2	44	59	59	68	72	72	69	72	72	55	73	73
Urban	1	42	48	48	32	35	35	33	33	33	35	38	38
Elevation (m)													
<250	1	26	32	31	32	33	30	35	34	30	32	35	37
250-500	2	27	38	36	64	58	49	64	55	46	69	68	66
500-750	3	81	74	69	96	91	86	96	89	78	102	100	99
>750	5	175	168	155	167	150	159	153	134	120	177	154	158
Cloud Cover (%)													
0-30	5	211	243	234	649	561	384	763	734	638	731	778	739
30-40	7.5	238	274	263	730	631	432	858	825	717	822	875	831
40-50	10	264	304	293	812	701	480	954	917	797	913	973	924
50-55	20	453	543	534	913	789	540	1073	1032	897	1028	1094	1039
55-60	30	643	782	775	1014	876	600	1192	1146	996	1142	1216	1155
60-65	37.5	723	880	871	1218	1135	966	1341	1290	1121	1284	1368	1299
65-70	45	803	978	968	1421	1394	1332	1490	1433	1245	1427	1520	1443
70-75	47.5	1004	1222	1210	1761	1832	1678	1863	1791	1556	1786	1914	1987
75-80	50	1130	1375	1362	1981	2061	1888	2096	2015	1751	2010	2153	2235
>80	55	1255	1528	1513	2201	2290	2098	2329	2239	1946	2233	2392	2484
R. Humidity (%)													
0-40	1	28	24	23	96	89	79	138	123	98	149	160	165
40-50	2	43	36	35	145	134	119	208	185	147	224	240	247
50-55	5	100	109	105	181	168	149	260	231	184	280	300	309
55-60	10	156	182	175	217	201	178	312	277	220	336	360	371
60-65	12	213	255	246	271	252	223	389	347	275	420	450	464
65-70	15	270	328	316	326	302	268	467	416	330	504	540	557
70-75	20	352	425	417	486	488	449	584	520	413	630	675	696
75-80	30	433	522	519	646	673	630	701	624	496	756	810	835
>80	35	650	783	778	933	888	843	1184	1141	994	1135	1215	1253

Land use	Weight in %	September 30 Years	September 20 Years	September 10 Years	October 30 Years	October 20 Years	October 10 Years	November 30 Years	November 20 Years	November 10 Years	December 30 Years	December 20 Years	December 10 Years
Forest	5	152	132	132	144	149	149	40	41	41	14	8	8
Shrub	4	110	110	110	118	123	123	25	21	21	8	6	6
Crop	3	89	86	86	85	102	102	15	19	19	8	6	6
Marshy	2	57	62	62	73	81	81	13	17	17	7	5	5
Urban	1	31	32	32	37	45	45	8	9	9	3	2	2
Elevation (m)													
<250	1	28	30	30	34	39	40	8	8	9	3	2	1
250-500	2	66	56	51	44	49	55	10	10	11	4	3	1
500-750	3	93	72	69	71	75	83	15	14	16	8	4	2
>750	5	149	120	124	147	146	157	44	32	35	15	5	3
Cloud Cover (%)													
0-30	5	745	642	548	246	276	293	39	39	41	14	10	5
30-40	7.5	838	722	617	277	310	330	44	44	46	15	12	6
40-50	10	931	802	685	308	345	367	49	49	51	17	13	6
50-55	20	1048	902	771	599	613	642	55	55	57	19	15	7
55-60	30	1164	1002	857	890	881	918	61	62	64	21	16	8
60-65	37.5	1226	1086	1005	1001	991	1033	68	69	72	24	18	9
65-70	45	1288	1169	1153	1112	1101	1148	76	77	79	27	20	10
70-75	47.5	1563	1739	1766	1390	1376	1435	95	96	99	33	25	12
75-80	50	1759	1957	1986	1564	1548	1614	107	108	112	37	28	14
>80	55	1954	2174	2207	1737	1720	1793	119	120	124	42	32	15
R. Humidity (%)													
0-40	1	145	117	103	78	87	101	23	23	24	8	6	3
40-50	2	218	175	155	117	131	152	35	35	36	13	9	5

50-55	5	272	219	193	147	164	189	44	44	45	16	12	6
55-60	10	327	263	232	176	197	227	52	52	54	19	14	7
60-65	12	408	328	290	220	246	284	65	65	68	24	17	8
65-70	15	490	394	348	264	295	341	78	79	81	28	21	10
70-75	20	613	493	435	443	493	536	79	89	96	29	31	15
75-80	30	735	591	522	622	692	731	80	100	112	30	41	19
>80	35	979	987	991	932	1038	1096	120	149	167	43	61	28

Bibliography

Ahmed R (1989) Probabilistic estimates of rainfall extremes in Bangladesh during the pre-monsoon season, Indian Geographical Journal, **64**:39-53.

Anderson James R (1971) Land use classification schemes used in selected recent geographic applications of remote sensing, Photogrammetric Engineering, **37** (4):379-387.

Anderson James R (1976) A land use and land cover classification system for use with remote sensor data: Geological Survey Professional Paper: 964.

Anderson James R, Hardy Ernest E, and Roach John T (1972) A land-use classification system for use with remote-sensor data: U.S. Geological Survey Circular, 671.

Andersson T, Mattisson I (1991) A Field test of thermometer screens, Swedish Meteorological and Hydrological Institute, Report No.: RMK 62, pp. 41.

Anon (1991) Methodology of priority demarcation survey, Department of Agriculture and Co-operation, Ministry of Agriculture, AIS & LUS Technical Bulletin no. 9.

Arnell N W (1999) A simple water balance model for the simulation of stream flow over a large geographic domain, Journal of Hydrology, **217**:314–335.

Bishop C M (1995) Neural networks for pattern recognition, Oxford: Oxford University Press.

Booth T H and Jones P G (1998) Identifying climatically suitable areas for growing particular trees in Latin America, Forest Ecology and Management, **108**:167–173.

Bouman B A M, Keulen H van, and Rabbinge R (1996) The 'School of de Wit' crop growth simulation models, A pedigree and historical overview, Agricultural Systems, **52** (2-3):171-198.

Bounoua L and Krishnamurti T N (1993) Influence of soil moisture on the Sahelian Climate Prediction I, Meteorology and Atmospheric Physics, 52(3-4):183-203.

Bruse M and Skinner C J (1999) Rooftop greening and local climate: a case study in Melbourne, Poster In: Proceedings International Conference on Urban Climatology & International Congress of Biometeorology, Sydney, 8 to12 November, Australia.

Burrough P A (1986) Principles of geographical information systems for land resources assessment, Oxford Oxfordshire and New York, 12, pp. 194-198.

Burrough P A and Donnell Mc (1998) Principles of geographical information systems, New York: Oxford University Press.

Changnon S A, Kunkel K E (1999) Rapidly expanding uses of climate data and information in agriculture and water resources: causes and characteristics of new applications, Bulletin American Meteorological Society, 80:821–830.

Chaplot V, Darboux F, Bourennane H, Leguedois S, Silvera N, and Phachomphon K (2006) Accuracy of interpolation techniques for the derivation of digital elevation models in relation to landform types and data density, Geomorphology, 77:126-141.

Collins F C and Bolstad PV (1996) A comparison of spatial interpolation techniques in temperature estimation. In: Proceedings of the third international conference/workshop on integrating GIS and environmental modeling, Santa Fe, New Mexico, January 21-25, 1996. Santa Barbara, California: National Center for Geographic Information Analysis (NCGIA).

Coltelli M, Fornaro G, Franceschetti G, Lanari R, Migiaccio M, Moreira J R, Papathanassaou K P, Puglisi G, Riccio D, and Schwabisch M (1996) SIR-C/X-SAR multifrequency multipass interferometry: A new tool for geological interpretation, Journal of Geophysical Research, 101:27-48.

Cramer W and Fischer A (1996) Data requirements for global terrestrial ecosystem modelling, In: Walker B, Steffen W (eds) global change and terrestrial ecosystems, Cambridge University Press, Cambridge, MA, pp. 530–565.

Daly C, Gibson W P, Taylor G H, Johnson G L, and Pateris P (2002) A knowledge-based approach to the statistical mapping of climate, Climate Research, **22** (2):99-113.

Das L and Lohar D (2005) Construction of climate change scenarios for a tropical monsoon region,. Climate Research, **30**:39–52.

Das T K, Das Gupta I V, Lohar D and Bhattacharya B (Eds.) (2011), Disasters in West Bengal – An interdisciplinary study, ACB Publications, Kolkata, pp. 23-27.

Dobson M C, Ulaby F T, Pierce L E, Sharik T L, Bergen K M, Kellndorfer J, Kendra J R, Li E, Lin Y C, Nashashibi A, Sarabandi K L, and Siqueira P (1995) Estimation of forest biomass characteristics in Northern Michigan with SIR-C/X-SAR data, IEEE Trans, Geoscience Remote Sensing, **33**:877-894.

Douglass D H, Pearson B D, and Singer S F (2004) Altitude Dependence of Atmospheric Temperature Trends, Geophysical Research Letters, **31**:1-4, doi:10.1029/2004GL020103.

Dowding S, Kuuskivi T, and LI X (2004) Void fill of SRTM elevation data – principles, processes and performance, In: images to decisions: remote sensing foundations for GIS applications, ASPRS, Fall Conference, September 12-16, Kansas City, MO, USA.

Du Q, Chang Ni-Bin, Yang C and Srilakshmi K R (2008) Combination of multispectral remote sensing, variable rate technology and environmental modelling for citrus pest management, Journal of Environmental Management, **86**:14–26

Dunn J A (1941) The origin of banded hematite ores in India, Economic Geology, **36**:355-370.

Easterling D R et al. (1997) Maximum and minimum temperature trends for the globe, Science, **277**:364-367.

Eckstein B A (1989) Evaluation of spline and weighted average interpolation algorithms, Computers and Geoscience, **15**:79-94.

Eischeid J K, Diaz H F, Bradley R S, and Jones P D (1991) A comprehensive precipitation data set for global land areas, US Department of Energy, Report No. DOEER69017TH1, Washington DC, pp. 81.

Elhance A P (1999) Hodropolitics in the Third World: conflict and cooperation in international river basin, US Institute of Peace Press, pp. 156–158.

Eliasson I (1992) Infrared thermography and urban temperature patterns, International Journal of Remote Sensing, **13**(5):869–879.

Eliasson I and Svensson M K (2003) Spatial air temperature variations and urban land use - a statistical approach, Meteorological Applications, **10** (2):135-149.

Fisher P F and Tate N J (2006) Causes and consequences of error in digital elevation models, Progress in Physical Geography, **30:** 467-489.

Fraser E (2008) Crop yield and climate change, Retrieved on 2009-06-14, http://www.enotes.com/topic/Climate_change_and_agriculture.

Gamache M (2004a) Free and low cost datasets for international mountain cartography, Documentation for the Alpine Mapping Guild, pp. 42.

Giorgi F and Francisco R (2000) Uncertainties in regional climate change prediction: a regional analysis of ensemble simulations with the HadCM2 coupled AOGCM, Climate Dynamics, **16**:169–182.

Hartkamp A D, Beur K D, Stein A, and White J W (1999) Interpolation techniques for climate variables, NRG-GIS Series 99-01, Proceedings of a Workshop, 19-22 April Mexico, CIMMYT, pp. 1-75.

Henderson S, Holman S, and Mortensen L (1993) Copyright information - modified with permission from global climates - past, present, and future, EPA Report U.S., Environmental Protection Agency, Office of Research and Development, Washington, DC, pp. 1-6.

Hengl T and Evans I S (2007) Geomorphometry: a brief guide, in 9 geomorphometry: concepts, software, applications, T. Hengl and H. I. Reuter 10 (Eds.), pp. 3-18.

Hulme M and Jenkins G J (1998) Climate change scenarios for the UK: scientific report, Climatic Research Unit, Norwich, UK, pp. 88.

Hulme M, Mitchel J F B, Jenkins J, Gregory J M, New M, and Viner D (1999) Global climate scenarios for fast-track impacts studies, Glob Env Change Suppl Iss:S3–S19.

Huqiang Z, Gao X, and and Li Y (2006) Impact of Land Use in China on regional climate: An Australia-China bilateral project on climate change, Proceedings of 8 ICSHMO, Brazil, April 24-28, pp. 955-957.

Hutchinson M F (1988) Calculation of hydrologically sound digital elevation models, Paper presented at Third International Symposium on Spatial Data Handling at Sydney, Australia.

Hutchinson M F (1989) A new procedure for gridding elevation and stream line data with automatic removal of spurious pits, Journal of Hydrology, 106: 211-232.

Hutchinson M F and Gessler P E (1994) Splines-more than just a smooth interpolator, Geoderma, 62:45-67.

Jackson Li F, Kustas T J, W P, Schmugge J, French A N, Cosh M H, and Bindlish R (2004) Deriving land surface temperature from Landsat 5 and 7 during SMEX02/SMACEX, Remote Sensing of Environment, 92:521-534.

Jain S K, Agarwal P K, and Singh V P (2007) Hydrology and water resources of India, Springer, pp. 5–9

Jarvis C H and Stuart N (2001) A comparison between strategies for interpolating maximum and minimum daily air temperatures: b. The interaction between guiding variable and interpolation method, Journal of Applied Meteorology, 40:1075-1084.

Jensen J R (1996) Introductory digital image processing: A remote sensing perspective, Second Edition, Prentice Hall, Upper Saddle River, New Jersey, pp. 318.

Jensen J R (2005a) Thematic information extraction: pattern recognition-A remote sensing perspective, Prentice Hall Series in geographic information science, Series Editor-Keith C. Clarke. 3rd Edition, Chapter 9, pp. 337-406.

Jensen J R (2005b) Thematic map accuracy assessment, introductory digital image processing- A remote sensing perspective, Prentice Hall Series in Geographic Information Science, Series Editor- Keith C. Clarke, 3rd Edition, Chapter 13, pp. 495-515.

Jingyong Z, Wenjie D, Lingyun W U, Jiangfeng W, Peiyan C, and LEE D K (2005) Impact of Land Use Changes on Surface Warming in China, Advance in Atmospheric Science, **22** (3):343–348

Jones P D (1994) Hemispheric surface air temperature variability-a reanalysis and update to 1993, Journal of Climate, **7**:1794–1802.

Jonsson P (2004) Vegetation as an urban climate control in the subtropical city of Gaborone, Botswana, International Journal of Climatology, **24**:1307–1322.

Karl T R, Williams C N, and Young P J (1986) A model to estimate the time of observation bias associated with monthly mean maximum, minimum and mean temperatures for the United States, Journal of Climate and Applied Meteorology, **25**:145–160.

Klink K and Willmott C J (1994) Influence of soil moisture and surface roughness heterogeneity on modeled climate, Climate Research, **5**:104-118.

Kottek M, Grieser J, Beck C, Rudolf B, and Rubel F (2006) World map of the Koppen-Geiger climate classification updated, Meteorologische Zeitschrift, **15**:259-263.

Krishnan M S (1968) Geology of India and Burma, Higginbothams (P) Ltd., Chennai.

Lillesand T M and Kiefer R W (2000) Remote Sensing and Image Interpretation, 4th ed. Wiley & Sons.

Longley P A, Goodchild M F, Maguire D J and Rhind D W (2005) Geographic Information Systems and Science, 2nd ed, Wiley & Sons.

Mallawaarachchi T, Walker P A, Young M D, Smyth R E, Lynch H S, and Dudgeon G (1996) GIS based integrated modelling systems for natural resource management, Agricultural Systems, **50**(2):169-189.

Matheron G (1970) The theory of regionalized variables and its applications, Issue 5, Les Cahiers du Centre de Morphologie Mathématique de Fontainebleau, Paris: École Nationale Supérieure des Mine, pp. 212.

Menz G (1997) Regionalization of precipitation models in East Africa using Meteosat data, International Journal of Climatology, **17**(10):1011–1027

Miles D L (1999) Estimating Soil Moisture, Crop series, irrigation, no. 4.700, Colorado State University Extension, **9**, pp. 98.

Miquel N, Pons X, and Roure J M (2000) A methodological approach of climatological modelling of temperature and precipitation through GIS techniques, International Journal of Climatology, **20**:1823–1841.

Mitchel T D and Jones P D (2005) An improved method of constructing a database of monthly climate observation and associated high-resolution grid, International Journal of Climatology, **25**:693–712.

Niyogi D (1970) Geological Background of Beach Erosion at Digha, West Bengal, Bulletin of the Geological Mining and Metallurgical Society of India, **43**, pp.1-36.

Oke T R (1982) The energetic basis of the urban heat island, Quarterly journal of the Royal Meteorological Society, **108**:1-24.

Pal D K, Sengupta S, and Dutta B (1994) Some special terrain features evaluation with ERS-1 SAR data in conjunction with IRS-1B LISS-II data, In Proceedings National Symposium on Microwave Remote Sensing & Users' Meet, 10-11 January 1994, pp. 220-223.

Pal D K, Venkataramana I, Sudhakar S, and Krishnan N (1992) Land use/ land cover mapping using IRS-1A data, Natural Resources Management - a new perspective, National Natural Resources Management System, Bangalore:360-366.

Pan Z, Arritt R W, Gutowski W J, and Takle E S (2001) Soil moisture in a regional climate model, Geophysical Research Letters, 1999GL000000, 0094-8276/01/1999GL000000$05.00.

Parton W J (1984) Predicting soil temperatures in a short grass steppe, Soil Science, **138**:93-101.

Pascoe E H (1973) A manual of the Geology of India and Burma, Controller of Publication, Govt. of India, New Delhi.

Pidwirny M (2006) Introduction to Soils, Fundamentals of Physical Geography, 2nd Edition.

Pielkel R A and Avissar R (1990) Influence of landscape structure on local and regional climate, Landscape Ecology, **4**(2/3):133-155.

Rao YP (1981) The climate of the Indian subcontinent, In World Survey of Climatology. K. Takahashi and H. Arakawa (Eds.), Elsevier, pp. 67-182.

Reuter H I, Nelson A, and Jarvis A (2007) An evaluation of void filling interpolation methods for SRTM data, International Journal of Geographic Information Science, **21**(9):983-1008.

Rigol J P, Jarvis C H, and Stuart N (2001) Artificial neural networks as a tool for spatial interpolation, International Journal of Geographical Information Science, **15**(4):323-343.

Ritter Michael E (2006) The Physical Environment: An Introduction to Physical Geography, http://www4.uwsp.edu/geo/faculty/ritter/geog101/textbook/title_page.html.

Samanta S (2009) Assessment of surface temperature using remote sensing technology, Papua New Guinea Journal of Research, Science and Technology, Papua New Guinea, **1**:12-18.

Servilla M and Towner M (2000) Integrating High-Resolution Satellite Imagery and Weather Data for Improved Agricultural Management Decisions, Proceedings, ESRI International User,Conference,http://proceedings.esri.com/library/userconf/proc00/professional/papers/P AP601/p601.htm.

Singh O P, Masood Ali Khan T and Rahman Md S (2001) Has the frequency of intense tropical cyclones increased in the north Indian Ocean? Current Science, **80**(4), 575-580.

Snyder M A, Kueppers M L, Sloan L C, Cavan D C, Jin J, and Kanamaru H (2006) Regional climate effects of irrigation and Urbanization in the Western United State: a model intercomparison, California Energy Commission, Public Interest Research Program,pp1-43.

Sobrino J A and Jimenez-Munoz J C (2005) Land surface temperature retrieval from thermal infrared data: An assessment in the context of the surface processes and ecosystem changes through response analysis (SPECTRA) mission, Journal of Geophysical Research, 110, doi: 10.1029/2004JD005588.

Sobrino J A, Jimenez-Munoz J C, and Paolini L (2004) Land surface temperature retrieval from LANDSAT TM 5, Remote Sensing of Environment, **90**:434-440.

Sobrino J A, Raissouni N, and Li Z (2001) A comparative study of land surface emissivity retrival from NOAA data, Remote Sensing of Environment, **75**:256– 266.

Spronken-Smith R A and Oke T R (1999) Scale modelling of nocturnal cooling in urban parks, Boundary-Layer Meteorol, **93**:287–312.

Stohlgren T J, Chase T N, Pielke R A, Kittel T G F, and Baron J S (1998) Evidence that local land use practices influence regional climate, vegetation and stream pattern in adjacent natural areas, Global Change Biology, **4**:495-504.

Tucker M R and Sear C B (2001) A comparison of Meteosat rainfall estimation techniques in Kenya, Meteorology Application, **8**:107–117

Upmanis H and Chen D (1999) Influence of geographical factors and meteorological variables on nocturnal urban-park temperature differences—a case study of summer 1995 in Goteborg, Sweden, Climate Research, **13**(2):125-139.

USGS (2006b) Shuttle Radar Topography Mission Level 1 (3-arc second) documentation, available online at http://edc.usgs.gov/products/elevation/srtmdted.html (accessed 01/05/2009).

Van De Griend, A A, and Owe M (1993) On the relationship between thermal emissivity and the normalized difference vegetation index for natural surfaces, International Journal of Remote Sensing, **14** (6):1119-1131.

Vogt J, Viau A, and Paquet F (1997) Mapping regional air temperature fields using satellite-derived surface skin temperatures, International Journal of Climatology, **17**:1559-1579.

Watson D F, and Philip G M (1985) A refinement of inverse distance weighted interpolation, Geo-Processing, **2**:315- 327.

Weng Q (2001) A remote sensing–GIS evaluation of urban expansion and its impact on surface temperature in the Zhujiang Delta, China, International Journal of Remote Sensing, **22**(10):1999-2014.

Yeh T C, Wetherland R T, and Manabe S (1984) The effect of soil moisture on the short-term climate and hydrology change- a numerical experiment, Monthly Weather Review, **112**:474-490.

Yokobori T and Ohta S (2009) Effect of land cover on air temperatures involved in the development of an intra-urban heat island, Climate Research, **39**:61–73.

www.ingramcontent.com/pod-product-compliance
Lightning Source LLC
Chambersburg PA
CBHW040903180526
45159CB00010BA/2914